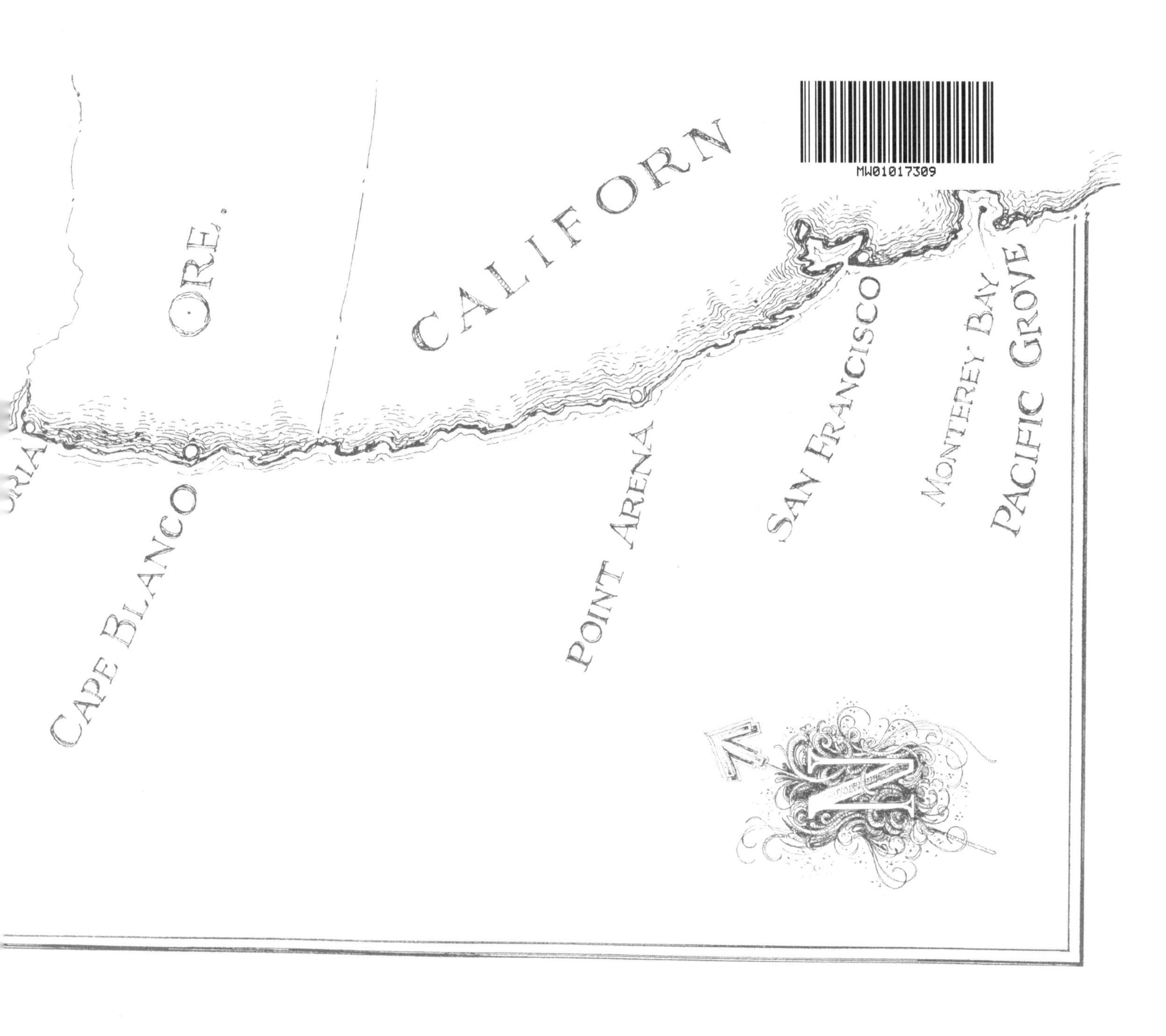
MW01017309
ORE.
CALIFORN
CAPE BLANCO
POINT ARENA
SAN FRANCISCO
MONTEREY BAY
PACIFIC GROVE
N

Ed Ricketts
from Cannery Row to Sitka, Alaska

ED RICKETTS
from Cannery Row to Sitka, Alaska

SCIENCE, HISTORY, AND REFLECTIONS
ALONG THE PACIFIC COAST

A Compilation of Essays edited by Janice M. Straley

REVISED EDITION

Old Sitka Rocks Press

Revised Edition © 2020

Published by Old Sitka Rocks Press
P.O. Box 273, Sitka, Alaska 99835
oldsitkarockspress@gmail.com

Library of Congress Control Number: 2020907928
ISBN: 978-0-578-68272-3

Printed in Canada

Cover Painting: Norman Campbell
Back Cover Photo: Ed Ricketts by Fred Strong, 1936
(Pat Hathaway Collection, CV# 95-033-001)

Book Design

This book is dedicated to the Renaissance men and women of Cannery Row and Sitka, Alaska.

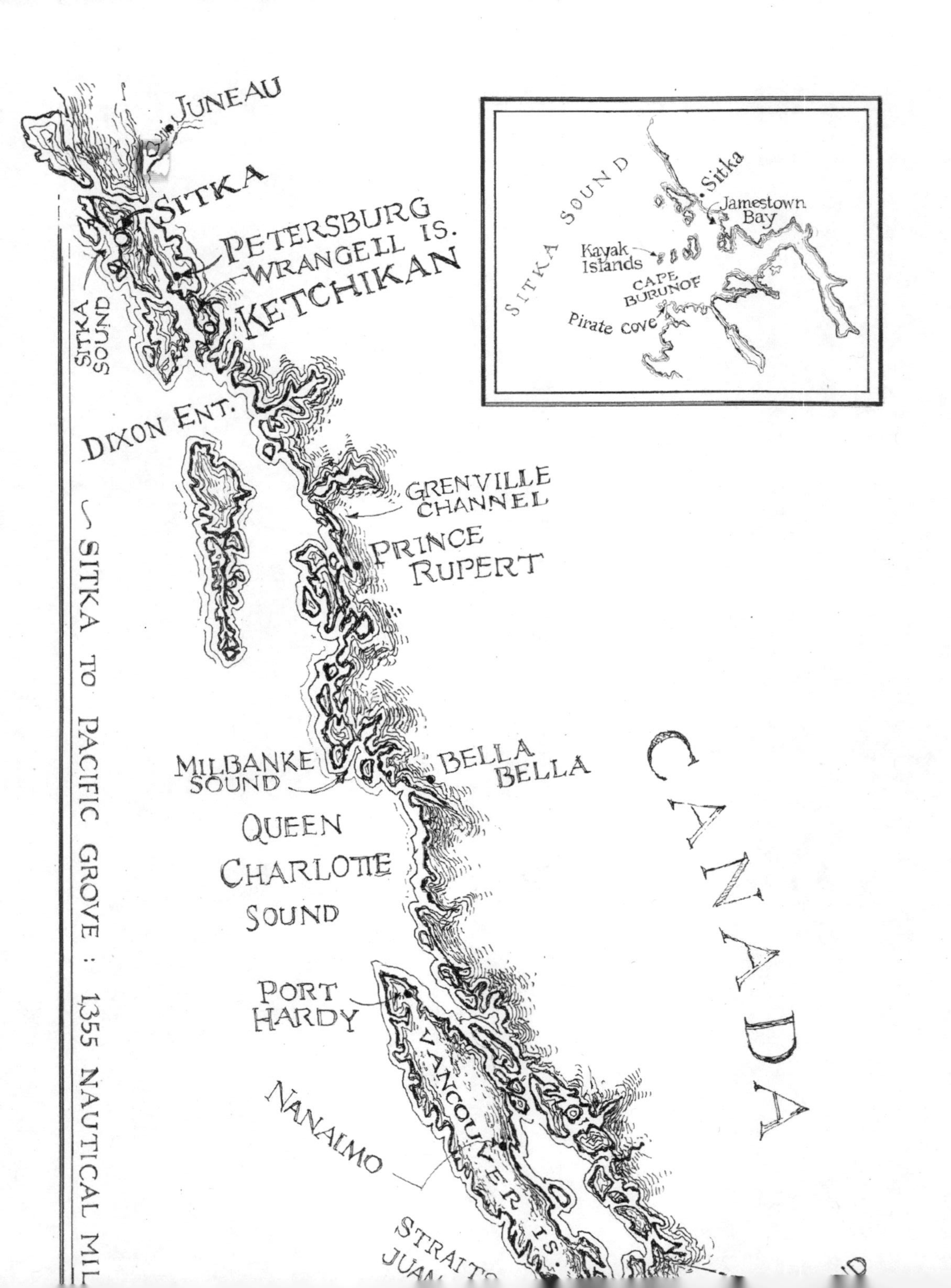

JUNEAU
SITKA
PETERSBURG
WRANGELL IS.
KETCHIKAN
SITKA SOUND
SITKA SOUND
Sitka
Jamestown Bay
Kayak Islands
CAPE BURUNOF
Pirate Cove
DIXON ENT.
GRENVILLE CHANNEL
PRINCE RUPERT
MILBANKE SOUND
BELLA BELLA
QUEEN CHARLOTTE SOUND
CANADA
PORT HARDY
VANCOUVER IS.
NANAIMO
SITKA TO PACIFIC GROVE : 1355 NAUTICAL MIL

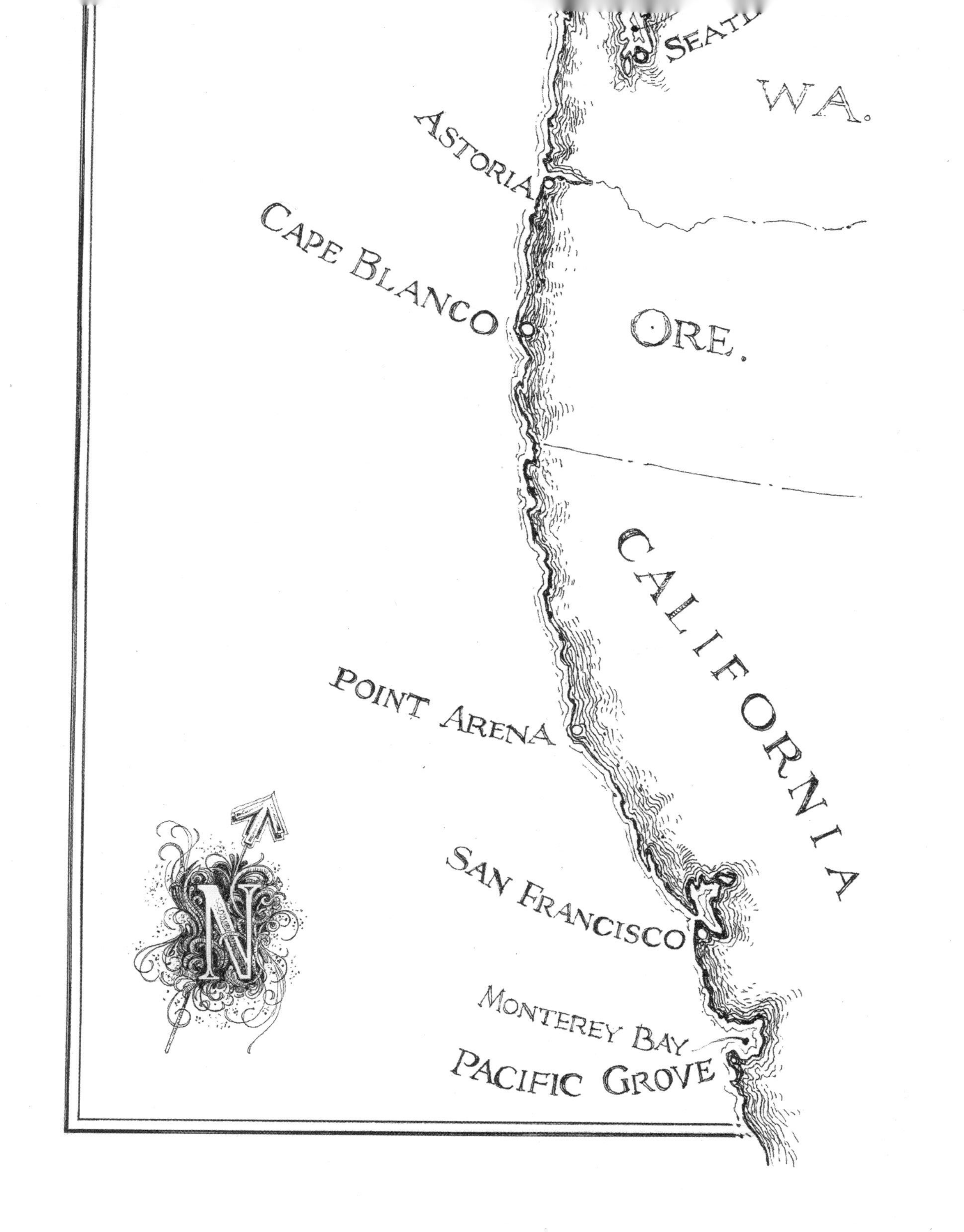
SEAT
WA.
ASTORIA
CAPE BLANCO
ORE.
CALIFORNIA
POINT ARENA
SAN FRANCISCO
MONTEREY BAY
PACIFIC GROVE
N

Contents

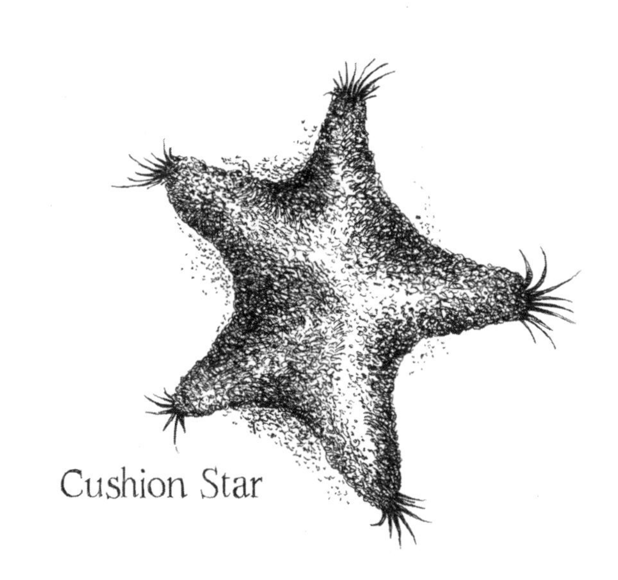
Cushion Star

Foreword

I was happy with this book when it came out in 2015, and am even more impressed with it now. It expresses many different views of a man who embodied even more of those views, all of them valid. That's what seems so satisfying: the views in this book are only a small part of Ed Ricketts, but all true.

I am not a scientist, although interested in science because of my early training. I am more focused on the humanities, the love of which I inherited from Dad. I believe brother Ed inherited more of Dad as a scientist than I did.

There may have been a time when we were a normal family, but I don't remember it. But we were surely a solid unit when we were collecting and on the trips that led to collecting. Dad was always on top of those trips. He drove the whole way as Mother didn't drive. Mother was in charge of keeping peace in the car. She was a marvelous cook and manager. We ate most of our lunches in wonderful surroundings—streams, meadows. Delicious food cooked over a can of Sterno; hand-cranked ice cream with fresh peaches or other fruits.

When we got to our destination, a cabin in Hoodsport, on Hood Canal in Washington state, we settled in. The next morning, early, early, Dad took over. We collected tirelessly 'til the tide came up and our little *Gonionemus* jelly fish were covered completely. We were all cold and exhausted. Mother cooked breakfast and Dad took care of the specimens. Ed and I occupied ourselves wherever we were needed, then played.

Brother Ed and I were very close during our early years, but after the family split-up we were separated for about seventeen years. Ed was in the Army in the Pacific during the time I was back on the Peninsula, and he returned from Japan at the end of the war, shortly after I was married and went east to Baltimore. When we saw each other again we filled many hours with talk about our lives. He kept track of events with Dad and I wrote of the years before. Between the two of us we must have a pretty good record of events in the Ricketts family. We also wrote to each other during the intervening years. He kept electronic records and I kept papers.

Dad had a humorous, fun-loving side. He loved words and wordplay. He loved telling stories to us, even made-up ones; he taught us long words, scientific names of tidepool animals. He loved music and loved to sing with other people; he knew songs from Lead Belly to Gregorian chant. Poetry, Marx Brothers, movies, reading, discussions, walking. Not a good mechanic, carpenter, fix-it person (but Mother was).

You will read several views of the same events and people in this book. We hope that you will come away with knowledge of the great wealth of scenery of the West Coast, especially Southeast Alaska, the arts and the philosophy of the people involved, and the magnificent world of life in the tidepools here. We are truly "Striving still to truth unknown"...

Nancy Ricketts
Sitka, Alaska | 2020

Preface

It has been over four years since the authors of *Ed Ricketts: From Cannery Row to Sitka, Alaska* gathered around my dining room table in Sitka to read and unify our essays. The first printing of this book sold in three years with very little marketing—a testament to the interest in Ricketts's life and writing, even years after his death.

There are a few changes in this revised edition. Artist Norman Campbell created a lovely new cover with tidepools and adjusted other drawings associated with the essays. A new essay has been included—"Ricketts, Calvin, and *Between Pacific Tides*" by Richard Astro and Don Kohrs—which links the voyage on the *Grampus* to the writing of *Between Pacific Tides* and the influence of that voyage on Joseph Campbell. The science essay, "The Legacy of a Naturalist" by Miner et al., was updated, and "Ricketts, Calvin, and the Charisma of Place" by John Straley was slightly revised.

This book of essays provides insight into how and why lives change in predictable and unexpected ways. The central theme is that place shapes your life. But it is also a story about observation and seeing the world within our own tidepool. John Steinbeck's foreword in the 1948 revised edition of *Between Pacific Tides* is a wonderful essay on observation and re-observation throughout time and how this propels us to be curious and to ask probing questions. Ed Ricketts saw the world through the lens of an intertidal pool. Steinbeck saw this world through the eyes of Ed Ricketts. Jack Calvin sought refuge and clarity in the wild natural world of Alaska.

We all see the world in our own unique ways. The "Wave Shock Essay" included in this book is clearly a trip through Ed Ricketts's brain waves. It is circular and no one idea or thought is cohesive, which is exactly why I love this essay. It is not a polished piece of writing but gives us such an intimate glimpse into how Ricketts's mind worked.

Nancy Ricketts is also a portal into her father's world. Her mind is sharp, as is her wit and sense of humor. At 95, she is a deep thinker, and her mind must be very much like her father's. I cherish the moments I have with her, though with each passing week she becomes more frail. She loves this book because it tells a story that encompasses who Ed Ricketts really was to her: a man who loved life and loved his children. As a child, she believed John Steinbeck never liked Ed's children. He portrayed Ed as "Doc" in his novels—a man with no family. This truly annoyed Nancy because this was not who her father was to most of the people who knew him. In this book Nancy can tell the world who Ed was as a father to her with her delightful essay about the family collecting biological specimens on the Olympic Peninsula. For this revised edition, I asked Nancy to write a foreword with additional details of their family life. She surprised me by graciously accepting my offer.

In cleaning and purging our home recently, I found my final high school research paper, written for Advanced Marine Biology in the spring of 1971 in Seattle when I was seventeen. I had completely forgotten about this report and was surprised to find it focused on wave shock and factors that influence how and where an organism lives. In this paper, I explained how the force from powerful waves can move whole communities and showed I did not really understand the theory of wave shock. For me, the most telling aspect of this paper was in reading the references which, while not numerous (there were three), are connected to my life today. I relied heavily on my beloved 1962 edition of *Between Pacific Tides,* and cited *Basic Ecology* written by Ralph and Mildred Buchsbaum. Ralph's photograph of Ed holding the jumbo squid at the Lab in 1936 appears in *Ed Ricketts: From Cannery Row to Sitka, Alaska* (page 14). Ralph and Mildred were the parents of Vicki Pearse, a noted invertebrate biologist, married to John Pearse, a co-author of the science essay in this book. Vicki and John live in the wonderful, curious house where Ralph had his textbook printing business, right across from Hopkins Marine Station and with a view of the beach where Ed was photographed sitting on the big rock in the 1920s, also included in this book (page 20). We have stayed with John and Vicki in their home and their basement is a treasure trove of books on the natural history of everything, from spiders to birds to snakes to anemones. I could have stayed browsing in that basement for days. In some ways, my life has come full circle, from my early fascination with marine ecology and my first discovery of Ed Ricketts, to the present as a whale biologist and editor of this collection of essays about him.

I believe the writings in *Ed Ricketts: From Cannery Row to Sitka, Alaska* can help guide us through a tumultuous world. Staying focused on the important factors that shape our lives—realizing that some influences are beyond our control but knowing others are within our grasp—can help to guide us to make the world a better place.

Jan Straley
Old Sitka Rocks | February 2020

INTRODUCTION: FINDING THE WORLD IN A TIDEPOOL

Janice M. Straley

"Along the surf swept open coast, the rocky cliffs of the ineptly named Pacific have developed associations of animals with phenomenal staying power and endurance of wave shock."
— Ed Ricketts and Jack Calvin, *Between Pacific Tides,* 1939

This collection of essays tells a remarkable story, connected through the central figure of Ed Ricketts, who gained worldwide recognition as John Steinbeck's fictional main character in the novel *Cannery Row*. Ricketts also achieved scientific notoriety as co-author (along with Jack Calvin) of the classic exploration of intertidal life, *Between Pacific Tides*.

The main essay in this collection was written by Ed Ricketts in 1932 and is published here for the first time. It explores the influence of wave shock upon littoral ecology, or in other words, how the impact from different wave sizes affects where animals live on the shore.

The original title of the essay, while clumsy, is impressively descriptive: "Notes and observations, mostly ecological, resulting from northern Pacific collecting trips chiefly in southeastern Alaska, with special reference to wave shock as a factor in littoral ecology."

The "Wave Shock Essay," as we refer to it in this book, introduces much of the early thinking explored in *Between Pacific Tides*. In one passage, Ricketts describes the similarities between the coastlines of Sitka, Alaska and the Monterey Peninsula in California:

> *The fauna of the surf swept rocks outside Sitka resembles that of the similarly exposed California coast nearly 2000 miles distant, more than it does that of similar type of bottom protected from surf, only three miles away.*

For Ricketts, this was evidence that exposure to wave impact determines what animals and plants will occur in a given place.

In the essay, Ricketts also establishes Sitka as a premier site for the study of intertidal biology, describing it as

> *the finest collecting place I have ever seen [...] Sitka provides a really excellent text book illustration for ecology. A short trip of 3 or 4 miles from the Sitka inner harbor to the fringing Kyack [sic] Islands or to Whale Island puts one into a new and more brilliant littoral world.*

Clearly, the coastline near Sitka was recognized by Ricketts as a significant location for furthering our knowledge of ecology.

The "Wave Shock Essay" revolves around a collecting trip through the Inside Passage from Puget Sound to Sitka aboard the *Grampus*, a thirty-three-foot vessel owned by Jack and Sasha Calvin. The remaining six essays in this book also have connections to Sitka and to Pacific Grove, California, where Ed Ricketts's lab was located. Together these essays tell the story of a group of people whose lives were intertwined with the ocean and with each other. The story details what they were thinking, feeling, and observing about the natural world and how the animals living along this coast survived in a brutal, powerful, and challenging seascape.

During the 1932 sampling trip aboard the *Grampus*, Jack Calvin and Ed Ricketts continued their progress on a book written for intertidal enthusiasts, describing the life in a tidepool. Originally conceived to be a handy pocket guide to intertidal creatures, the book had evolved into a broader, more comprehensive accounting of the intertidal life from California to Alaska. The resulting text, *Between Pacific Tides*, was a hard sell to Stanford University Press whose editors believed the organization of animals should be based upon taxonomy —a system of scientific classification—rather than how an animal lived its life in a tidepool. In *Between Pacific Tides*, Ricketts and Calvin used an ecological perspective to classify the tidepool inhabitants; animals who lived in similar physical and biological environments were classified together.

It took years for the first printing of *Between Pacific Tides* to occur, but it was eventually published in 1939. Although the reasons for the delay may not be completely known to anyone except Ricketts and Calvin, there was a long drawn-out editing process, perhaps stemming from the editors' perception that Ricketts and Calvin lacked the proper credentials to author such a book. In today's thinking, Ricketts had a holistic approach to science and to life, bringing ideas and people together. These ideas came from the study of music, literature, and poetry, in addition to the study of science. I imagine Ricketts always must have had more to say to his companions than what was in a tidepool, during the late night collecting trips for his biological supply business, or after a day of exploration along the shore on the *Grampus*.

But Ricketts was also a scientist to the core and he was fully engaged within the scientific community. He collected and sent preserved animals to researchers across continents, seeking the correct species identification and scientific name. He compared notes with scientists on the details of the habitat characteristics for these animals as well. The list of scientific papers, contacts, and references cited in both the "Wave Shock Essay" and *Between Pacific Tides* rivals the best scientific writers for thoroughness and accuracy.

The ivory tower academics from Stanford could have been critical of the way Ricketts describes certain species as having human traits and behaviors (e.g. "takes in boarders" in reference to a species which hosts parasites). In the world of science, making such human parallels with the animal world was, and still is, unacceptable. This was perhaps another strike against Ricketts by the editors, but this aspect of his thinking brings all things full circle, making it clear humans cannot be completely objective about an ecological study and cannot be removed from the larger picture.

To admit subjectivity within your scientific view is rarely, if ever, expressed in western science. However, the intertwining of the human and animal worlds is commonplace within indigenous cultures. In this respect, Ed Ricketts was thinking about the connections among people and animals in a different way than his western science colleagues.

The first two editions of *Between Pacific Tides* extensively featured the area around Sitka, and the ideas developed within the "Wave Shock Essay." In these editions the subtitle specifically connected Sitka to the coastline farther south: "An Account of the Habits and Habitats of Some Five Hundred of the Common, Conspicuous Seashore Invertebrates of the Pacific Coast between Sitka, Alaska and Northern Mexico." Later editions included a map on the inside cover but only depicted the area from Baja, Mexico to Vancouver Island, Canada. By the fifth edition, published in 1985, the science of intertidal biology had grown significantly and new scientific information was added and some chapters were deleted. Sitka was featured less prominently in these later editions.

Between Pacific Tides was a seminal book—a landmark in marine ecology conceived by an observant collector of intertidal species who was a vividly descriptive writer. It was initially intended to be sold for educational purposes. However, what the editors missed was the passion and dedication of Ed Ricketts, which made readers excited about looking at a tidepool. This passion exudes from *Between Pacific Tides*, especially in the early editions.

Humor, detailed observations and critical thinking made reading *Between Pacific Tides* fun; at least it was for me. As a junior in high school, with a love for the ocean that stemmed from growing up exploring the coast of Washington State and the San Juan Islands, I fell in love with this book and the ideas it contained. When offered a job in Sitka as a young adult, newly married to a farrier and aspiring writer whose landscape was the arid country of eastern Washington, I was really in a dilemma. How could I miss this opportunity?

All I knew about Sitka was from what I read in *Between Pacific Tides*. Sitka was where all the known range extensions ended for many tidepool invertebrates. Fortunately, my husband was up for the adventure, and we moved to Sitka for seasonal jobs—and ended up building our lives together in Alaska. I eventually became a whale researcher and science educator, and I can honestly say that *Between Pacific Tides* shaped my life.

I think Ed Ricketts may have seen parallels between the community of animals living within a tidepool and a community of people traveling in a small boat. Each community has a combined energy, and changes—however minute—can alter the behavior of the animals or people within their respective closed environments. In particular, back in 1932, describing how wave shock can alter a community of organisms was a new way of observing from an ecological perspective. The ideas and concepts presented in the "Wave Shock Essay" helped forge a new paradigm for the way we understand how intertidal communities develop and how the animals behave with increasing exposure to waves pounding the coastline.

Ed Ricketts was formulating these concepts when he tragically died in 1948, leaving many of his ideas unfinished. The "Wave Shock Essay" may not have been ready for publication from Ricketts's perspective because his thoughts on the subject were in a state of constant flux.

However, the "Wave Shock Essay" does provide an enlightening view of how ideas germinate and evolve within an engaged mind. In her essay, "Ed Ricketts and the Ecology of a Science Writer," Katharine Rodger—the preeminent biographer of Ricketts—eloquently brings him to life, capturing his gregarious personality and ability to captivate and engage an audience around a tidepool. She describes how writing was Ricketts's way of bringing together ideas about science and philosophy, particularly during his trip to Sitka.

Ed Ricketts's daughter, Nancy, moved to Sitka in 1974 with her husband who worked for the US Forest Service. She became a passionate archivist, working as a librarian for Sheldon Jackson College, and over the span of many years has been archiving her collection of her father's legacy. I got to know Nancy and her incredible attention to detail while finishing my graduate thesis in the late 1980s, before the age of the Internet. She was amazingly skilled at locating the most obscure literature references for me. Now in her early 90s, she has more energy and spunk than most people half her age.

For this project, Nancy allowed me to scan her entire collection of photographs, across the course of many months. The stories she told me about the people involved were invaluable in shaping this book. Her memories of her father, her aunt (Ed's sister Frances) and their family experiences were wonderful background music during the seemingly endless scanning process. Ed Ricketts was clearly a loving father and this is evident in the photographs and letters archived by Nancy. She brings this devotion to life in "Some Memories of the Early Days of the EFR Ricketts Family," a poignant essay describing the family collecting trips and their early years together in California prior to the *Grampus* trip.

The Calvins—Jack and Sasha—had known Ed Ricketts in Monterey, and in 1932 traveled with him aboard the *Grampus* from the Puget Sound area, surveying and collecting along the way to Sitka. Alaska author Colleen Mondor's essay describes life aboard the *Grampus*, bringing the reader to understand the relationships between an unusual mix of writers, readers, thinkers, and scientists who were together for most of the summer of 1932.

During this trip, three sampling sites in Sitka were intensively surveyed. These sites formed the basis for examining the theory of wave shock, the focus of Ricketts's essay. In 2012, two of these sites—Pirate Cove and the Kayak Islands—were re-established as biodiversity monitoring sites by the Sitka Sound Science Center. Data from surveys in 2012, as well as subsequent years (sites were resurveyed in 2018), can be use to document changes across the intervening years. The qualitative nature of collecting in 1932 was very different, however, from the highly quantitative methods used eighty years later, so results from the two survey approaches could not be directly compared to assess changes over time. A comprehensive list of the species identified for each site is included in this book, as well as the references for a report summarizing the 2012 study.

In partnership with marine ecologists at the University of California Santa Cruz, the Sitka sites are now included in a long-term monitoring program along the U.S. west coast. Survey results have shown the Kayak Islands to have species biodiversity that rivals the highest documented in biodiversity surveys along the British Columbia and Alaska coasts. This would not have been unexpected news to Ed Ricketts.

In their essay, "The Legacy of a Naturalist: How Ricketts's Wave Shock Idea Helped Shape a Century of Shoreline Research," marine scientists Melissa

Miner, Dave Lohse, Pete Raimondi, and John Pearse document the scientific contribution of Ricketts's ideas about intertidal ecology in the 1920s and 1930s and compare those ideas to the current thinking regarding marine communities. Their essay explores the "mostly ecological" past and present value of wave shock theory.

The book's final essay, "Ricketts, Calvin, and the Charisma of Place," is written by John Straley, novelist and former Alaska Writer Laureate. John evocatively describes the similarities between historical Monterey and current day Sitka, and offers a reflection on why place matters. This essay brings the ideas Ed Ricketts was thinking, discussing, and writing about full circle, from Sitka, Alaska to Cannery Row in California, and from the *Grampus* to the tidepool. Who you associate with and where you live will define how you live your life.

Taken together, these essays allow us to gain insight as to how Ed Ricketts thought and how he observed. Ricketts loved life to the fullest; he loved the written word, people, beautiful places, music, and the mysteries of the ocean. His curiosity for life provided an opportunity to make connections with people everywhere. Specifically, he brought remarkable thinkers and readers aboard the *Grampus* to observe the connections and relationships among the intertidal animals along this wave-swept coastline.

ED RICKETTS and THE ECOLOGY of a SCIENCE WRITER

Katharine A. Rodger

Ed Ricketts holding a jumbo squid. This photograph was taken in 1936 by Ralph Buchsbaum, author of the innovative textbook, *Animals Without Backbones*. The book includes photographs showing jellyfishes, corals, flatworms, spiders, and other invertebrates in their natural habitats.

Ed Ricketts has become, for many, a legendary figure. He was a marine scientist, a philosopher, a proto-ecologist, a hero, a womanizer, a saint, a renaissance man, a mystic, and a guru, depending on which account of him one encounters. His son, Ed Jr., once wisely noted that the sometimes contradicting portraits of his father reflect the various ways in which people came to know him, and that no one image was truly complete or accurate. Perhaps what fascinates people is the very complexity of Ricketts's life and interests; his ability to do so many things and to mean so much to those who knew him.

Virtually every written account of Ed Ricketts has tried—in varying degrees—to lay claim to presenting the truest version of him. Each biographer, friend, or scholar has offered a new and more definitive account of his life and work, often highlighting the heretofore "missing link" to fully understanding Ricketts in a person, place, or event that has been overlooked or underemphasized in previous portraits. Yet those of us who have presented these portraits have all fallen short—and many of us realize it, as we have felt the dissatisfaction of knowing that our words haven't fully captured the vibrancy of Ricketts's personality and relationships, that the limitations of language prevent us from describing the intensity of his intellect and passions.

Perhaps the best way to pay tribute to, and to learn from, and to understand more about Ed Ricketts, is to look more carefully at his own writings—to see the ways in which he learned to write about his life and work.

Edward Flanders Robb Ricketts was born in Chicago in 1897. His childhood was most notably shaped by spending time outdoors—even in his urban environment—but especially during trips outside the city limits. At the age of ten, he and his family moved to South Dakota for a period of one year, and some say it was there that his affinity for nature truly was born. Ricketts loved to camp and sleep outside, and according to accounts of his childhood by his sister, he was an avid reader who took an early interest in biology, thanks to the gift from an uncle of a zoology textbook. Ricketts's first experiences as a collector were of local insects and curiosities, including butterflies and bird eggs.

Ricketts's formal education was not particularly remarkable. He enrolled first in the Illinois State Normal University, and later at the University of Chicago, though he did not complete a degree at either. But at both schools, he studied science: three courses in zoology at Illinois State Normal University, and evolution, genetics, vertebrate paleontology, and animal ecology at the University of Chicago. His brief but formative encounter with the eminent zoologist, W.C. Allee, at the University of Chicago, with whom Ricketts studied animal ecology in the fall of 1922, was likely the final push that moved him west to Monterey Bay, one of the biologically richest spots in California. There, Ricketts, his wife Nan, and their infant son Ed Jr. began a life that would be closely intertwined with the marine ecology of the North American Pacific.

In their early years in Monterey, the Ricketts family grew to include two daughters, Nancy Jane and Cornelia (later called Bee). Together, they explored, observed, and collected specimens along the coastline for Ricketts's blossoming business, Pacific Biological Laboratories, a biological supply house that provided samples for schools across the country. While collecting trips took the family from San Diego to Vancouver, the Monterey Bay itself was rich in biological diversity and kept Ricketts fascinated with the ecology he found there. Nan Ricketts recalled the early years of collecting as marked by curiosity—not only their own, but also of the people that observed them in the tidepools, wondering what Ricketts and his family were doing.

In 1925 Ricketts drafted a catalog for Pacific Biological Laboratories that advertised his services and supplies. Of interest, however, is his description of the Monterey Bay itself and his caution about the risks of depleting it:

> *Monterey Bay is the fusion point of faunas from the North and South, and the ranges of a number of characteristic species of both regions overlap in these waters. [...] It should be borne in mind, (and this applies especially to local marine forms), that we must, above all else, avoid depleting the region by over collecting. One or more formerly rich regions, according to reliable authorities, already afford instances of the ease with which depletion is brought about.*

Though his notion here of a "fusion point" is geographical, the term itself describes the kind of nexus of Ricketts's own modus operandi. That is, in this early articulation of his work, Ricketts reveals that the ecological abundance of the region is what will make his business successful (though it never flourished financially), and yet that very abundance ultimately endangers it. This passage may represent one of the earliest and most prescient statements about the need for conservation that was published about the region.

Ricketts's interests in Monterey Bay's ecological diversity expanded beyond the needs of his business, and he began collecting marine animals that he did not recognize. As marine biologist and iconoclast Joel Hedgpeth notes, "He sent specimens to the Smithsonian and to other specialists for identification. He also observed the creatures' ways in the tidepools." As Ricketts began to focus on these animal behaviors and "ways" of interacting in the intertidal zone, he also began to share his ideas and observations. At first he discussed his ideas informally; Nan remembers him talking with folks in the tidepools while he worked. He had an engaging manner, and welcomed questions and people's interest. Soon friends and family encouraged him to write down what he knew in a kind of guidebook—an idea that he initially declined, but later became increasingly interested in pursuing.

One of the factors that likely pushed Ricketts to begin writing was his introduction to Jack Calvin in Monterey. Calvin was a writer and former English teacher who lived nearby with his wife Sasha. The men formed a friendship based on mutual interests and a love of exploring the natural world. According to Calvin, he suggested that Ricketts write about the local marine life, and by the mid to late 1920s, they had begun working together on a book about the topic. Calvin was a talented photographer and writer, and he enjoyed working with Ricketts in the tidepools. Ricketts, in turn, found in Calvin someone who could help articulate his ideas about how the physical environment impacted the species within it.

Because of a devastating fire in 1936 that destroyed virtually all of Ricketts's notes and correspondence, we do not have evidence of how, exactly, he and

Calvin worked together on drafting the book that would become *Between Pacific Tides*. Many have speculated about the extent to which Calvin may have co-authored, edited, or merely illustrated the book that derived out of Ricketts's work and thinking about ecology. But we know for certain that the men worked together. And without question, *Between Pacific Tides* reflects Ricketts's ideas about the factors that affected animal behavior and distribution—including wave shock, tides, and habitat.

As many have noted, this approach to science was relatively new and certainly unique to the existing body of research on the North American Pacific marine life in the early twentieth century. For those of us interested in how Ed Ricketts thought about the natural world, *Between Pacific Tides* is one of the earliest documents we have that describes that thinking. And perhaps most interesting is how his descriptions articulate a unique voice and style of writing that was not often found among scientists or within scientific writing at the time. Ricketts sought to make his writing accessible to the lay reader, and though his book includes ample citations and detailed data, it also includes rhetorically engaging descriptions that draw in readers, both lay and expert alike. Crabs were a particular favorite, and Ricketts's observations convey his enjoyment of his subjects as well as their behavior:

> *The pleasant and absurd hermit crabs are the clowns of the tide pools. They rush about on the floors and sides of the rock pools, withdrawing instantly into their borrowed or stolen shells at the least sign of danger, and so dropping to the bottom. [...] Among themselves, when they are not busy scavenging or love-making, the gregarious "hermits" fight with tire-*

CLOCKWISE FROM TOP: The lab at Pacific Biological Laboratories was located downstairs in the basement next to the garage. *(Photo by Fred Strong, 1936. Pat Hathaway Collection, CV# 95-033-004.)* Ed Ricketts on the beach at Hopkins Marine Station circa late 1920s. *(Photo by Ed Ricketts, Jr. Pat Hathaway Collection, CV# 81-021-0068.)* Charles Abbot Ricketts's car parked in front of the original lab on the Monterey waterfront. The upstairs living area was brought over from Pacific Grove and placed on top of the garage and lab. The courtyard outside had a staircase to an upstairs room where white rats, rattlesnakes, and Gila monsters were kept for research purposes. *(Courtesy of Nancy Ricketts.)*

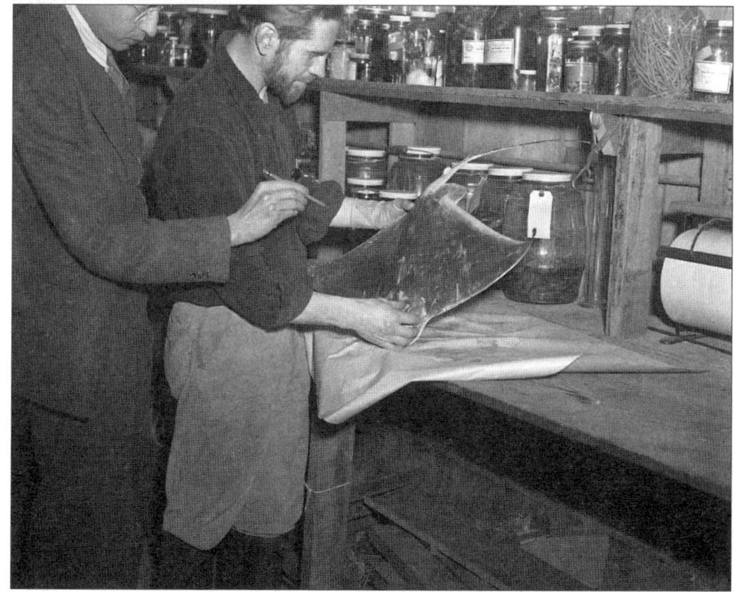

TOP: Charles Abbott Ricketts, Ed's father, embalming dogfish at Pacific Biological Laboratories. He moved from Chicago to Monterey with his wife Alice in the 1920s. *(Photo by Fred Strong, 1936. Pat Hathaway Collection, CV# 95-003-006.)*

BOTTOM: Ritchie Lovejoy and Ed Ricketts with a ray. *(Pat Hathaway Collection, CV# 81-021-070.)*

less enthusiasm tempered with caution. Despite the seeming viciousness of their battles, none, apparently, are ever injured. When the vanquished has been surprised or frightened into withdrawing his soft body from his shell, he is then allowed to dart back into it, or at least to snap his hind-quarters into the shell discarded by his conqueror.

This excerpt not only illustrates the sense of humor and enjoyment that Ricketts brought to his observations, but also the anthropomorphic connections he was quick to recognize in the tidepools. He did not see the behaviors of animals as unique or separate from those exhibited by humans; instead he found greater understanding of the ways in which our own actions and reactions might be impacted by our surroundings as well as by our peers. This particular idea derives directly from the work of Allee, though Ricketts was able to expand its application to specimens unique to the Pacific coast.

The originality and import of *Between Pacific Tides* are not isolated to quirky descriptions. The book is also among the most fully developed texts that Ricketts wrote in his lifetime.

Ricketts's life was cut short in 1948 when his car was struck by a train in Monterey. Though he survived the impact of the crash, he died days later at the age of fifty. The premature loss of Ed Ricketts can never be overstated—from the pain felt by his children, family, and friends, to the professional chasm left in his unfinished work. John Steinbeck's tribute to his friend expresses the immediate period of grief with poignancy. "A kind of anesthesia settled on the people who knew Ed Ricketts," he wrote.

There was not sorrow really but rather puzzled questions—what were we going to do? How can we rearrange our lives now? Everyone who knew him turned inward. It was a strange thing—quiet and strange. We were lost and could not find ourselves.

Ricketts left a number of projects behind that would have expanded and solidified his scientific legacy: a book on the outer shores of Vancouver Island and Southeast Alaska, a series of papers on plankton that presaged the el Niño phenomenon, and preliminary outlines for a book about California's desert regions, among others. But in spite of what was left undone, Ricketts's extant writings provide for us the portrait of a complex thinker, and also of a unique science writer.

Between Pacific Tides includes some of Ricketts's first statements about his philosophical interests as well as his scientific ideas. He spent considerable time and energy articulating complex ideas derived from Zen Buddhism, Taoism, and other philosophical traditions. His three essays—"Non-Teleological Thinking," "Breaking Through," and "A Spiritual Morphology of Poetry"—have been collectively considered the framework of his "unified field hypothesis," or personal philosophy. Yet Ricketts did not think—nor did he write—about things in isolation. That is, though he attempted to narrow his focus in each of those essays for the sake of clarifying his ideas about each particular subject, it would be a mistake to consider them as separate from his scientific thinking or writing, and vice versa.

Ed Jr., who lived at the lab after his parents separated, remembers his father working on projects night and day. "I would wake up and hear him at the typewriter," he says. "He was always working, always writing." Ricketts's philosophical essays were developed and constantly revised and reworked throughout the 1920s and 1930s, and evidence of the ideas within them can be found in *Between Pacific Tides*. In particular, Ricketts includes a few brief, but deliberate, connections between the phenomena he witnesses in the natural environment and his philosophical perspective. In describing brittle stars, for instance, "with long snaky arms bearing spines that take off at right angles," he notes that when aggregated under rocks, they are "so closely associated that their arms are intertwined." He goes on to suggest that these occurrences "lead us to the borderline of the metaphysical."

The notion of a "borderline of the metaphysical" itself was a significant idea to Ed Ricketts. In many of his writings, especially in his philosophical essays, he explores the idea of borders and boundaries—which he often called the "knife-edge." He sought places, both physical and metaphorical, where realities and worlds blurred, and where he might catch glimpses of something more or "beyond" what was in front of him. Philosophically speaking, he called these moments "breaking through" (a term he appropriated from Robinson Jeffers's poem, "Roan Stallion"), and he believed that such moments were more likely to be experienced if one practiced non-teleological, or "is," thinking. Poetry, music, literature, and art were all mediums through which an individual might find this transcendent border and see beyond. But so was nature itself. The tidepools were sites of delight and fascination for Ricketts, but also portals for profound understanding. These were the places where the knife-edge dissolved enough for him to see how things came together—where what he called his "unified field hypothesis," or world view, might be realized.

Among the most notable places and times during which Ricketts explored and further developed his unified field hypothesis was while traveling Alaska's Inside Passage in 1932—a trip that was essential in shaping *Between Pacific Tides*. This was a formative trip for Ricketts in many ways, garnering the attention of scholars and historians who have delved into his relationship with Joseph Campbell, who had become an intimate friend and collaborator in the months before their expedition. Ricketts and Campbell were also accompanied by Jack Calvin—along with Sasha Calvin—and without question, the men shared a deep interest in the environs they encountered as well as literature and philosophy. As noted in the work of both Eric Enno Tamm and Stephen and Robin Larsen, Ricketts and Campbell found in their daily exploration of the shores between Puget Sound and Southeast Alaska not just a scenic backdrop for their philosophical musings, but a literal representation of some of those ideas.

The collecting notes from the trip were assembled into a typescript called "Notes and observations, mostly ecological, resulting from northern Pacific collecting trips chiefly in southeastern Alaska, with special reference to wave shock as a factor in littoral ecology." Beyond the scientific value of Ricketts's detailed observations about species and terrain, the typescript as a whole represents a kind of tidepool in which he sees not just the marine life, but the entire ecosystem he encountered. Thus a typical description includes not only the scientific data and descriptors about observed specimens, but also a broader picture of the environment itself:

> *At Bella Bella Cannery, Indian Village, and store, where at least one old wood carver is doing excellent work on cedar chests, and where the morale of the*

Indians is high, if one may judge by their clean and interesting "lived-in" houses, there is a bit of low and rolling country, and some depositing shore. On July 14th specimens of the preposterous pelagic nudibranch Melibe leonina *were taken here in an eel grass cove. These are large (to 5 or 6"), fragile, transparent and delicate, but they swim well if grotesquely by snapping the body sideways. The hood, filled with air, is said to serve as a float, but the animal can also attach to such support as eel grass with some force, wrapping its foot around the stalk.*

The significance here is more than just scientific; this passage is representative of Ricketts's attempts to see the world holistically. In trying to understand the importance of an environmental factor that impacts marine life—i.e. wave shock—he knows that it is not only the "preposterous" nudibranch that is affected, but the entire coastal community, humans and animals alike.

Numerous other observations about the human towns and villages they encountered are found in the document as well, as he was equally fascinated by the microcosm of the tidepools and the macrocosm of the region. Thus his "Wave Shock Essay" is a rich account not only of how Ricketts came to understand wave shock more fully, but how he came to understand how the forces of the world we live in can shape us.

In spite of losing Ricketts too soon, and in spite of the fire that destroyed so much of his work, we are left with an impressive legacy of notes, essays, letters, and notebooks that say so much about his thinking and his life. But it would be a mistake to look at any one document or trip as the "most significant," and Ricketts himself would likely discourage us from trying to pinpoint what made him different. Instead we must look at the whole collection in an attempt to see the complexities that made him exceptional—but also that made him human. As he and Steinbeck wrote in *Sea of Cortez*, "It is advisable to look from the tidepool to the stars and then back to the tidepool again."

SOME MEMORIES of THE EARLY DAYS of the EFR RICKETTS FAMILY

Nancy Ricketts

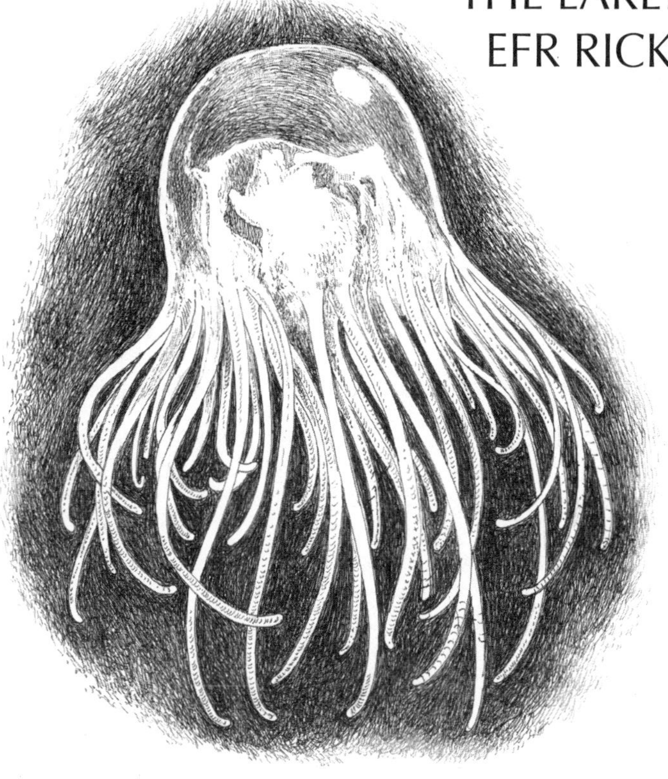

The Ricketts famly celebrating Ed Jr.'s 8th grade graduation. LEFT TO RIGHT: Ed's mother Alice, daughters Cornelia and Nancy, brother-in-law Fred Young, and Ed Jr., circa 1937. *(Courtesy of Nancy Ricketts.)*

Of the things I remember earliest, my favorite times were those in the living room of the 221 Fourth Street house in Pacific Grove. It must have been in the late twenties. Christmas was a special time, but it is Sundays I am really getting to. I don't know how often it happened, but often, I think. Dad and a friend or two sat in the living room and talked. There was the delicious smell of cigars, and you could see the smoke, with sun-rays shooting through it, circling to the ceiling. There were never raised voices, just quiet discussions on many topics.

Brother Ed and I, it seemed, were always in the living room, quite contentedly, and it seemed to me, quietly, just playing, and sometimes listening to conversation. I can't remember sister Cornelia at all at these times; she must have been a baby at the time, and in a crib—or with Mother? Mother spent a lot of time in the kitchen, and since she was a good cook and it was Sunday, there was always the smell of a roast in the house.

The living room was quite dark, but friendly-dark. The windows were fairly high and small, but quite beautifully designed. Ed and I enjoyed the chairs in the living room, although it doesn't seem like they could have been very comfortable; they were of a very hard, fairly light-colored wood and had black leather seats. I don't remember any couch, but the alcoves had built-in seating. We had two alcoves in the living room; one was high on the street side, with a cupola-type window, (brother Ed's favorite), the other was my favorite, with a window overlooking the front door on the side of the house. There was a gate-leg table on the inside wall facing the bedroom Cornelia and I shared. I think there were two bedrooms, a dressing room and a bathroom on one side of the house, forming one leg of the U-shaped house. The living room connected it with the other leg of the U which held the kitchen, a utility room, a maid's room and a bathroom. Inside the U was a covered patio, a lovely place that we all enjoyed.

The house was quite close to the street, but there were only a few windows on that side. The outside walls of the house next to the street were covered with a climbing rose—*Cecile Brunner*. From the street there was a walkway down the side of the house to a very graceful and beautiful entrance with a wide front porch and shallow steps of stone; the front door was flanked by full-length windows. The huge backyard carried over to the next street and was a real beauty with lots of oak trees, periwinkle, cineraria, and other colorful plants. No one in our family was hot on gardening and for a while we had a gardener. We never had a maid for the maid's room—that was used by Dad for a home lab. Cornelia wandered in there once (no one knew how she got in as it was off-limits to us) and drank something from a bottle. Dad gave her the antidote, which made her drunk enough that they had to watch that she didn't go too close to the stove. I can't remember where the stove was located. My golly, I can't even remember where the fireplace was, though we MUST have had one.

I can remember our first phone and a phonograph in that house. It was surely the beginning of my interest in music, as Dad played a lot of records.

Nicknames were quite popular with us—especially with Dad. I was "tata," Ed and I were "Sheik and Sheba." Ed was "boy," then "Junior," then "Edward," and later would answer to nothing but "Ed." For me there was "wormy" and "peaches" and "mugwumps" and "Nancy Jane, Butterfly Name." Cornelia was "Cornelia Frances N Connie." Sometime later when she was stung by a bee and ran to Dad for comfort, he told her next time to "bite him back;" her name became "Bit-a-bee," shortened eventually to "Bee;" I was calling her Bee well into our teens, and later she adopted "Rikki."

LEFT: Anna "Nan" Makar Ricketts, mother of Ed Jr., Nancy, and Cornelia, circa 1931. RIGHT: Ed Ricketts with children Ed Jr. and Nancy, circa 1928. Photo was taken at the 221 4th Street house. *(Courtesy of Nancy Ricketts.)*

One of brother Ed's and my favorite treats was having Dad read to us and tell us stories in the evening before bedtime. He also let me sit on his lap and braid his hair, comb down over his face, and other assorted goodies. Well, the reading—I can remember ONLY Homer's *Odyssey*, from a very thin, small book translated into Middle English. It was lost in the Lab fire in 1936. Dad would read only a short time each night, I suppose partly to hold our attention, and also because he had to answer so many of our questions: "Yes, but did Circe REALLY turn the men into pigs?" He told us a lot of stories also, mostly about animals, and they were TWIN animals, giraffes with names such as "Montgomery and Montmorency." There were also monkeys that were forever wetting on people.

We had turtles and lizards on our patio. The "floor" was made of flagstone, and the back "wall" was glass; the roof leaked periodically. We had salamanders in the yard and it was only years later that I heard they were called newts. We were familiar with horned toads, and made the acquaintance of tarantulas, Gila monsters, and rattlesnakes quite early.

Our summer trips to Puget Sound started possibly about 1930. Dad went to Alaska in 1932. When he returned he went down to Santa Barbara and brought the rest of the family back to Monterey Peninsula, at which time we relocated just "over the hill" in Carmel. Then there were more trips to Puget Sound.

The first trip I actually remember was from Carmel; we had a used tan Packard from Stahl's Automobile Dealership in Monterey. Special items for the trip were designed (and built?) by Mother—wooden structures to hold glass

LEFT: The home of the Ricketts family until 1932 at 221 4th Street, Pacific Grove. Joesph Campbell lived nearby.
BELOW: Ed Ricketts leaning against the Packard Limousine that was used during the 1930s for family collecting and surveying trips to the coast. A sliding window separated the front from the back of the car, where the kids (and chickens) sometimes rode. *(Courtesy of Nancy Ricketts.)*

specimen jars and other collecting equipment. Sometime later we got a wonderful, huge, formerly chauffeur-driven Packard (also from Stahl's), which allowed lots more packing space, as it had folding jump-seats in back, a wind-up window between the front seat and the back section, AND a microphone (seems like the microphone was deemed off-limits).

Packing was carefully done. (For a couple of days? It seemed like forever to us kids). Dad always drove, and planned to arrive in Hoodsport, Washington, for the first of the "low spring tides." We went directly north, stopping for gas and restrooms when needed; sometimes we thought we didn't have to "go" when we stopped, but after we had left the gas station, one of us often found we should have gone when we had the chance. One time that I vividly remember, we got going before brother Ed hopped in the car. Bee and I were crying, Mother wondered what all the fuss was about, and Ed was running after the car.

We very often stopped for lunch when we found a lovely stream, maybe even a gloriously wooded area with a clearing near it; we'd spot the place from the car. Mother often shopped along the road at vegetable and fruit stands and came up with such goodies as corn, tomatoes and peaches; she was always supplied with a can of Sterno and voilà—lunch. We carried a wooden ice cream freezer, and with some cream, salt, and plenty of shared muscle-power, we had wonderful ice cream.

Dad was fascinated with high tension electrical lines, wondering why they were at that location, and where they led—we followed them more than once.

We all liked old broken-down houses too. Dad was prone to high volume sneezing, which seemed hazardous to us, especially when sailing around zippy curves while we were in the Siskiyou Mountains between California and Oregon. More than once he pulled to the side of the road when he felt a sneeze coming on, and we encouraged that.

We kids entertained ourselves sometimes with waving to train engineers; they always waved back. We counted the number of railroad cars. We checked the license plates of cars to see what state they were from, and we even checked out the make of the car, as we knew them all by name, there being fewer companies then and a lot fewer models. But the most fun was checking out the Burma-Shave signs along the side of the road—little advertising signs for a shaving cream. One that I remember as being very clever was "Don't stick your elbow," (then wait for the next sign), "Out the window too far," (gap), "It might go home," (gap), "In another car."

At nighttime we usually stopped at an auto-court, but there were times when we used the back seat of the car plus sleeping bags. Was it only one night on the road? I don't remember. We got to our summer auto-camp in Hoodsport the night before the first low tide.

Always up at the crack of dawn, we hopped into bathing suits with a sweater over them and old tennis shoes (to protect feet from the vicious barnacles), had a quick hot breakfast (probably oatmeal), piled into the car and drove miles to our chosen collecting spot, aiming to get there just a little before the tide was low.

Out came the equipment—sieves, jars, flashlights for us kids and other equipment for Dad. Most of the time we were collecting *gonionemus*—a thumbnail sized (more or less) jellyfish found in the eel grass. We VERY carefully picked up the specimens in the sieve and lowered them into the jar filled with seawater. When we had a full jar, we took them to Dad, who placed them in a larger container. I guess it was pretty cold, but the serious business of carefully collecting kept us occupied until the tide came in to the point where the eelgrass was well-covered and no more jellyfish could be seen.

Back into the car, and back to the auto-court, and then the rest of the day started for us kids. Dad preserved specimens and Mother cooked. It was FUN! We loved it.

Sometimes during the summer we drove to Shelton for a movie; Groucho Marx was one of Dad's favorites, but we saw Greta Garbo, Leslie Howard, and others, and such movies as *The Scarlet Pimpernel*. One time Dad laughed so hard at one of the Groucho movies he fell off his seat and we practically had to disown him. Then we usually went to the Chinese restaurant. Chinese food was a family favorite; I can only remember fried rice! For a really BIG occasion we went as far as Olympia for dinner and a movie.

At the end of the "collecting season," we headed back for the Monterey peninsula, and school. Sometimes Mother collected exotic things to bring home, like banty chickens to provide us with eggs; these were kept in back with us kids for the whole trip. However, Mother kept the goose we got once on her lap in the front seat.

Seems like our family was slowly splitting up after the 1932 trip, but summers were the exception. Eventually Dad went to live at the Lab and the rest of us lived in Carmel. In about 1938, Mother, Bee and I headed north on a Greyhound bus, eventually ending up in Washington. Ed stayed with Dad. I was twelve or thirteen.

LEFT: Nancy Ricketts modeling the dress she made for a USO dance during World War II. RIGHT: Ed. Jr. was an amateur musician, circa 1940, in Monterey. His father, with whom he lived during his teenage years, thought he might play professionally one day. *(Courtesy of Nancy Ricketts.)*

LATER

Fast-forward to the early 1970s. I had moved up to Sitka with my Forest Service husband when our combined family had grown up and left, and I was working at Sheldon Jackson College's Stratton Library. I was also taking a few college courses that appealed to me. When the good ones ran out I took a number of independent studies, sometimes with instructors from different places. I was engaged in one that called for comparisons between Tlingit myths and world-wide myths, and one of my sources was a Joseph Campbell book. I suddenly came up with "Joseph Campbell, Joseph Campbell—could that possibly be Joe Campbell from my childhood on Monterey Peninsula?" I wrote to Campbell in care of his publisher and much later, in September of 1981, I got a handwritten letter postmarked New York, from Joe.

> *What a delightful surprise, to receive a letter from Nancy Ricketts—the little girl I waved goodbye to in 1932: and from Sitka, Alaska, no less, where her father and I had such a wonderful season of fishing for thousands of little jellyfish, a bit later that year… I count your father one of the most important and best beloved friends of my lifetime. His death has not taken him away from me…*

THE *GRAMPUS*

Colleen Mondor

Joseph Campbell (third from left), Sasha Calvin (in doorway), and Ed Ricketts (far right) with *Grampus* and unidentified visitors, 1932. *(Photographer unknown, Pat Hathaway Collection, CV# 99-026-006.)*

As the motorboat *Grampus* sailed out of Puget Sound on June 29, 1932, the four people onboard expected nothing more than an unremarkable collecting and surveying trip of the 1,100 or so miles of coastline between Tacoma, Washington and Juneau, Alaska. Known by mariners as the Inside Passage, it was an area intimately familiar to the boat's owners, Jack and Sasha Calvin, who had traveled there in 1929 on their honeymoon.

Describing that earlier trip in a 1933 article for *National Geographic* magazine, Jack Calvin wrote that he and Sasha found the very notion of such an expedition to be irresistible. The names on the map were a lure that Calvin invited the magazine's readers to ponder:

> *Who could resist the delightful agnosticism of No Use Ledge politely questioning Yes Bay? What dabbler in philosophy could not meditate for hours on the waters of Celestial Reef or off Khayyam Point? And a confirmed wanderer could but find incentive for farther wanderings in Danger Point, Misery Island, Turnagain Island and Stop Island.*

With their dog along for company, the Calvins paddled six to ten hours a day in their seventeen-foot cedar canoe. The *Nakwasina* was built by the legendary Willits Brothers Canoe Company of Tacoma and included a sail, although they were only able to utilize it for a very limited period. The trip took fifty-three days, most of them spent with the couple enduring steady rain. Navigating by often outdated and inadequate charts, and paddling close to the shore for reference, the Calvins accomplished with comparative ease what many people along the way described as foolhardy, if not downright dangerous. Calvin took great pains to point out that he and his wife experienced no such concerns, although a single stint of night paddling was so cold and unpleasant that they resolved themselves to daylight activity only.

The biggest problem for the couple actually was would-be rescuers who could not imagine the Calvins were in the canoe voluntarily. More than once, they were nearly sunk by the unwanted efforts of those who sought to "save" them. Their safe and timely arrival in Juneau thus heralded an achievement over both man and weather, and when the opportunity to conduct a collecting trip along a similar route with Ed Ricketts was proposed, they were more than up to the challenge.

The voyage of the *Grampus* was funded by a commercial contract—with Ed Ricketts's Pacific Biological Laboratories—to collect 15,000 specimens of the small pink jellyfish, *Gonionemus vertens*. Ricketts also wanted to "make a hurried ecological reconnaissance of the shore invertebrates" and to collect some of what he termed "ecological horizon markers," in an effort to better understand the broader coastal area, species and influences.

Ricketts invited his friend Joseph Campbell to come along on the trip, and the writer, who had been struggling through several years of unemployment, quickly agreed. Although they had met only a few months earlier when Campbell arrived on the west coast looking for work, the two men felt a strong sense

Grampus, Ed Ricketts crouching on dock, 1932. *(Photographer unknown, Pat Hathaway Collection, CV# 99-026-005.)*

of kinship. In an interview with the *New York Times* decades later, Campbell explained the impact Ricketts had on his life and work:

> *Talking with Ricketts, I realized that between mythology and biology there is a very close association. I think of mythology as a function of biology; it's a production of the human imagination, which is moved by the energies of the organs of the body operating against each other. These are the same in human beings all over the world and this is the basis for the archetypology of myth. So, I've thought of myself as a kind of marginal scientist studying the phenomenology of the human body, you might say.*

Although Campbell might have only been looking for a bit of adventure and some time for literary inspiration (he brought along Oswald Spengler's *The Decline of the Western Empire*, and Fyodor Dostoyevsky's *The Idiot*, among other books, to read and discuss), the *Grampus* journey would end up being, in his words, "one of the primary personal transformations of a life dedicated to self-discovery." While Ed Ricketts was finding different ways to study and understand the creatures of the northern Pacific Ocean, Joe Campbell was discovering a unique way of looking at the world and the universal stories that compel us to live.

The thirty-three-foot long *Grampus* had basic accommodations: the Calvins bunked in the forward cabin; Ricketts slept on a bench in the aft cabin and Campbell slept beside him on the floor. The *Nakwasina* was secured on top of the *Grampus,* and Ricketts and Campbell regularly took the canoe out at low tide on collecting trips. Ricketts would also disappear for hours on his own, walking along the shore in the early mornings and returning, as Campbell noted, with "his bags full of curious creatures, over which they would all marvel."

There was a degree of serendipity in the way the *Grampus* expedition had come together so easily. Ricketts and the Calvins were part of the same close group of Pacific Grove friends which included John and Carol Steinbeck; Sasha Calvin's sister, Tal Lovejoy; Tal's husband, Ritchie; and Joseph Campbell. The friendships within the group were complicated and intense, highlighted by boisterous gatherings and deeply thoughtful conversations.

All members of the group were supportive of Ricketts's ongoing efforts to transform his Pacific Biological Laboratories' specimen sales catalogs into what he envisioned as an informal textbook of Pacific Coast invertebrates. As the book project expanded, Jack Calvin became co-author and contributed photographs, Ritchie Lovejoy provided drawings, and scientists from around the world identified Ricketts's specimens. *Between Pacific Tides* was under consideration by Stanford University Press in 1932, before the *Grampus* departed, but publication was delayed after a devastating review by Dr. W.K. Fisher of Stanford's Hopkins Marine Station.

In his assessment of the manuscript, Fisher emphasized Calvin and Ricketts's lack of formal scientific credentials, insisting that "the manuscript should be carefully read by a professional zoologist. It must be remembered that neither of the authors can be classified in this category, although Mr. Ricketts is a collector of considerable experience."

On the *Grampus* voyage, Ricketts continued to revise his manuscript in an effort to address some of Fisher's concerns. He was determined to craft something that appealed to both the layman and to zoologists. As biologist Joel Hedgepeth would later note in an updated and revised edition of *Between Pacific Tides*, the book was "essentially a guide for those who do not dive but who approach the seashore in boots or old tennis shoes." Ricketts intended it to be a book for people like himself and his friends—those who came to the tidepools with wonder and curiosity. The challenge for him was to find a way to satisfy the academics while keeping his vision intact for the common beachcomber. Thus, the book still needed work and he kept at it on the *Grampus*. Joseph Campbell would eventually read it while the group was in Sitka and offer some editing suggestions.

The Calvins handled daily operations for the *Grampus* while slowly navigating the route north. That was especially tricky as the only available charts largely still reflected information from outdated surveys. As Ricketts noted, "The main steamer routes only have been adequately charted." One of the charts even carried the disclaimer that it should not be relied upon for navigation.

Jack Calvin maintained a straightforward captain's log throughout the *Grampus* trip, noting weather, stops for repairs or refueling, and places where Ricketts was particularly successful in hauling in his jellyfish and other sea creatures. Campbell wrote more deeply about his perceptions of the region and his developing thoughts about wilderness and civilization, biology and myth—ideas which were sparked by the long conversations onboard the

Grampus. Ricketts used a notebook that recorded both scientific data in a manner similar to Calvin's navigational entries and more far-reaching ruminations on ecology and the environment. That original notebook is presumed to have been lost in the 1936 fire at Pacific Marine Laboratories. The only portions of it that remain are those found in the "Wave Shock Essay."

The *Grampus* traveled north by what Ricketts described as "early stages, traversing the inland channels via Nanaimo, Pender Harbor, Alert Bay, Prince Rupert, Ketchikan, Wrangell, Petersburg and Sitka." They stopped and collected whenever possible, with Calvin and Ricketts always on the lookout for *Gonionemus vertens* and also for areas that might prove the most fruitful for finding other specimens to expand Ricketts's research into local marine life.

The group spent three weeks in Sitka, canoeing, collecting, walking and philosophizing. Ricketts typed up his collecting notes. Campbell read his books and recorded his own observations about the landscape around them, the culture of the Native peoples they had encountered along the way and the vast differences between the lives they knew in California and this new one they were experiencing in Alaska.

Always, Ricketts studied and took note of what he saw as they traveled, what he found on his collecting forays, and his conclusions about the region's shoreline and how it related to its ecological development. The "Wave Shock Essay" is the result of those careful observations and an indication of the type of work he hoped to continue in future journeys to Alaska.

The *Grampus* trip lasted ten weeks, ending in Juneau with a visit to Sasha Calvin's family, including her father, Andrew Petrovich Kashevaroff, the Russian Orthodox priest at the town's St. Nicholas Church. On August 26 Ricketts and Campbell boarded the *Princess Louise* and departed for Seattle. It would be the last trip they took together.

The Calvins, who at that point had already made Sitka their home, would go on to become powerful voices in the area's conservation movement. Jack and Sasha, along with others, would establish the Sitka Conservation Society which later spearheaded efforts to establish the West Chichagof-Yakobi Wilderness there—the first citizen-initiated Wilderness Area in the United States.

Joseph Campbell returned to the east coast in September 1932, and ultimately achieved literary fame with his explorations of myth and storytelling. Several of his books became bestsellers, including *The Hero With a Thousand Faces* which, along with the television series *The Power of Myth*, helped to extend his popularity far beyond academic circles. He had planned to write a novel or screenplay about his time in Monterey, tentatively titled "The *Grampus* Adventure," and also, according to his biographers, spoke after Ricketts's death "of a dramatic work in which he compared Ed Ricketts's life to the classic structure of the life of the Hero." However, neither of those projects were ever written for publication.

Ed Ricketts learned enough during his collecting forays while on the *Grampus* to write the "Wave Shock Essay," but he believed there was far more to the story of the Inside Passage and the North Pacific than one scientific paper could explore. That bigger project, of the entire North American Pacific coastline, was something he continued to work on as he completed *Between Pacific Tides* and resubmitted it to Stanford in 1936. In 1939, the year after it was finally published, Ricketts and John Steinbeck began a six week expedition to the Gulf of California that would be recounted in *Sea of Cortez: A Leisurely Journal of Travel and Research*—co-authored by Ricketts and Steinbeck, and published in December 1941.

For Ricketts, those two books were part of a greater plan, critical components of what he now envisioned as a trilogy on the Pacific. He outlined his goals in a letter to Stanford University Press on February 11, 1942:

> *My study of the marine invertebrates has divided itself geographically into three divisions. The first, on the animals of the US and Canada, appeared in* [Between Pacific Tides]. *A second, the appendix to the Sea of Cortez, extended the range to the south....The third part...will deal with the Aleutians, Gulf of Alaska, Bering Sea, etc.*

That last book was never written. Ed Ricketts died before he could conduct the necessary research and thus the voyage of the *Grampus* serves largely as a tantalizing glimpse into what might have been. In the "Wave Shock Essay," Ricketts referred to Sitka as "the finest collecting place I have ever seen." But his 1932 Alaska adventure had a far more significant impact than the record of marine specimens he found there. Onboard the *Grampus*, Ricketts, Joseph Campbell, and Jack and Sasha Calvin brought their combined intellectual curiosity together with a fierce desire to experience the world in its wildest state. It was not enough for them to stand on the shore and look out at the ocean—they wanted to go out in its depths and see what they could find there. This was a small but fascinating group, brought together for a brief time, who left the slightest of records behind to recall their ten weeks together on the water. But the "Wave Shock Essay," the photographs from the trip, and their scattered reflections in logbook entries and diary pages are enough to remind us of what they accomplished, and more importantly, why they were all so determined to set out on the Pacific in the first place.

RICKETTS, CALVIN, and *BETWEEN PACIFIC TIDES*

Richard Astro and
Donald Kohrs

Stanford University's Hopkins Marine Station, 1920. The two women in the picture are Myrtle E. Johnson and Gertrude Peirson. The man is quite possibly Harry James Snook. Johnson and Snook were authors of *Seashore Animals of the Pacific Coast*. *(Photographer: Walter K. Fisher, 1920. Photograph courtesy of Stanford University Archives.)*

It is an unfortunate irony that Ed Ricketts was, in life and for more than a quarter century after his death, best known as a character in the fiction of Nobel Prize winning novelist, John Steinbeck. Even after his untimely death in a car-train accident in May of 1948, we knew Ricketts best as Doc Burton in *In Dubious Battle*, as Preacher Casy in *The Grapes of Wrath*, as Doc Winter in *The Moon Is Down*, and as Friend Ed in *Burning Bright*. But Ed Ricketts was neither a physician (*In Dubious Battle*), a preacher (*The Grapes of Wrath*), a Scandinavian doctor (*The Moon Is Down*), nor a mystic (*Burning Bright*). Closer to the truth, but still subject to authorial license, are Steinbeck's portraits of him as Doc in *Cannery Row* and *Sweet Thursday*, as a laboratory scientist in "The Snake," and in Steinbeck's tribute to him in an afterward to the narrative portion of *Sea of Cortez*. Even more unfortunate was the portrait of Ed in David Ward's 1982 film, *Cannery Row*, in which Nick Nolte plays Doc/Ed as a quirky recluse with a shadowy past as a professional baseball pitcher who nearly killed a batter with his fastball.

Over the years and particularly during the half century after his death, Ricketts became a cult figure, a sort of Jerry Garcia of American marine science. Initially, the Ricketts cult—they called themselves "Ed Heads"—were Monterey area locals who would sit on the steps of his former laboratory and reflect on "the hour of the pearl" on Cannery Row. And then, "Ed Heads" emerged elsewhere—in marine laboratories along the Pacific Coast and among readers of *Cannery Row* who made pilgrimages to Monterey Bay to pay tribute to the man about whom Steinbeck wrote:

Some he taught how to think, others how to see or hear... Children on the beach he taught how to look for and find beautiful animals in worlds they had not suspected were there at all.

"Ricketts is like a cult figure," remarked Joseph Taylor, a professor of environmental history at Simon Fraser University. Of all the scientists who ran marine research stations along the Pacific Coast, Taylor says, "he was by far the most colorful." We've been told that scientists at the University of Victoria on Vancouver Island have erected a small shrine to Ricketts that includes his portrait as well as a glass-enclosed cabinet crammed with pickled seashore specimens he collected.

Thanks in large measure to Katie Rodger who, with Ed Ricketts, Jr., edited two important books about Ricketts's work and his person, a handful of serious scholars have recognized that Ricketts was not simply a character in fiction, but a scientist whose biologically-based worldview helped inform the thematic design of Steinbeck's most important novels. Those insights, though quite valuable, were gleaned almost entirely from *Sea of Cortez* (1941), the record of the Ricketts-Steinbeck expedition to the Gulf of California in the spring of 1940.

Sea of Cortez contains much of Ricketts's thinking about marine biology—a habitat approach to life on the seashore that places him stage-center in the then-emerging field of marine ecology. What has been overlooked, however, is that the scientific and philosophical insights in *Sea of Cortez* have their origins in *Between Pacific Tides*, Ricketts's pathbreaking study of marine life along the central Pacific coast that he authored with Jack Calvin, and that features drawings by Ritch Lovejoy. Moreover, Ricketts's many insights about science generally, and marine life specifically, that he and Steinbeck garnered aboard the *Western Flyer* in the Gulf of California, can be found in part or in full in the work Ricketts did with Jack Calvin, his wife Sasha, and budding mythologist Joseph Campbell during a ten-week expedition through Alaskan waters during the summer of 1932. When we review what we know of the conversations between Ricketts and Campbell on that trip, we find that Ed's impact on Steinbeck's fiction was no greater than it was on Campbell's thinking—thinking that informs some of the most important works in comparative mythology in American intellectual history. In a paragraph about Ricketts that Steinbeck wrote in the Appendix to *Log from the Sea of Cortez*, the novelist affirms that "no one who knew him will deny the force and influence of Ed Ricketts. Everyone near him was influenced by him, deeply and permanently." Both Steinbeck and Campbell were among that number.

It was in April of 1930 when Jack Calvin first approached Stanford University Press proposing that he and Ricketts would write "a non-technical handbook for the casual seashore visitors (and for classes in nature study) illustrating and describing the common inter-tidal animals of the Pacific coast between Alaska and Lower California." In fact, by the time Calvin approached Stanford University Press, Ricketts was seven years into a career as a collector of biological specimens and the owner of his biological supply house, Pacific Biological Laboratories on Monterey's Cannery Row. When he arrived in California in 1923, Ricketts immediately began traveling the coastline on collecting trips to stock the supply house, sometimes with his wife and children and/or with resident and visiting scientists from Stanford's Hopkins Marine Station which was located around the corner from Ricketts's Pacific Biological Laboratories.

These trips allowed Ricketts to survey the rocky coasts and sand beaches of Monterey Bay, the mudflats of Elkhorn Slough, and the wharf pilings of Monterey and Santa Cruz harbors. With the help of local fishermen, he'd collect specimens from the nearshore pelagic and deep-water forms at the bottom of the bay. Venturing beyond Monterey Bay, he also visited and collected along the shores of Southern California (1925, 1929), Northern Mexico (1925), and Seattle/Puget Sound (1929). In early 1930, he traveled with scientists George and Nettie MacGinitie to the shores of Southern California and Mexico. That August, he went to Washington State with Jack and Sasha Calvin. In 1931, he repeated his trip south, this time with Torsten Gislén, a visiting scientist from Sweden. He repeated it again in the spring of 1932 with the MacGinities. All of these trips occurred before he boarded Jack and Sasha Calvin's thirty-three-foot boat, the *Grampus*, on a collecting trip from Tacoma to Sitka in June of 1932.

He spent a goodly portion of his time on these trips collecting invertebrates that he sold to high school and college biology departments. But his main interest was surveying the habits of marine invertebrates in their various habitats—rocky coasts, sand beaches, mud flats, and wharf pilings. This work came to define the habitat-based organization of *Between Pacific Tides*. Notebook and pencil in hand, Ricketts surveyed the topography of the shores where he collected, studying the environmental factors of wave shock, type of bottom, and tidal level.

As we know, Ricketts was first introduced to habitat approaches to the study of marine invertebrates by zoologists at the University of Chicago who were on the cutting edge of the emerging field of animal ecology. Ricketts studied at Chicago for a time, though he never graduated, contrary to what Steinbeck tells us in "About Ed Ricketts." There he encountered, among others, the pioneering ecologist Warder C. Allee, who impressed Ricketts greatly and whose writings were Ricketts's constant companions, starting with Allee's first major publication on the subject that appeared in 1923. In fact, the surveys Ricketts conducted and questions he asked ran parallel to those Allee conducted near Woods Hole and outlined in this published work. But Ricketts would not rely solely on Allee for his understanding of the intertidal zone. After relocating to Pacific Grove and establishing Pacific Biological Laboratories, Ricketts began connecting with professional zoologists (systematics and invertebrate specialists), and pioneering intertidal ecologists in the U.S. and around the globe, even though he was largely self-taught and had only short term exposure to academic science.

Ricketts's association with these scientists was symbiotic: they helped him identify the marine invertebrates he had collected, while Ricketts provided the scientists with specimens they could describe taxonomically in their own scientific publications. The Hopkins Marine Station served as an important conduit for Ricketts to established marine scientists, and enabled him to develop a working correspondence with such luminaries as Waldo L. Schmitt and Ira E. Cornwall of the U. S. National Museum of Natural History, Herbert Lyman Clark of Harvard's Museum of Comparative Zoology, Libbie H. Hyman and Willard G. Van Name of New York's American Museum of Natural History, and E.S. Pilsbry of the Peabody Academy of Science in Philadelphia. At Hopkins, Ricketts established collaborative relationships with Director Walter K. Fisher, Willis G. Hewatt, George and Nettie MacGinitie, and Max W. De Laubenfels.

Ricketts's work with scientists at the Hopkins Marine Station and elsewhere, along with his collecting trips up and down the Pacific coast, was ideal preparation for his expedition with the Calvins and Joseph Campbell on the *Grampus*. At thirty-three feet, the *Grampus* was not a large vessel, and while the waterways they traversed were largely sheltered from the open ocean, there were difficult crossings with tricky currents and hidden shoals. In fact, it was not until Ricketts convinced Campbell that the Calvins were experienced sea travelers—proof positive of which was their honeymoon trip over the same waters in a nineteen-foot canoe three years earlier—that Campbell agreed to sign on. But sign on he did, and as he, Ricketts, and the Calvins moved north, they were struck by the differences in coastal habitats and, in particular, by the differing effects of wave shock on the distribution of life in the intertidal. Ricketts wrote of their collecting for several weeks in localities protected from any significant wave energy:

> *We had collected for weeks in an environment free from ground swell and surf. Then suddenly, within a few miles, both appeared, we were again on open coast. And more than coincidentally, the whole nature of the animal communities changed radically, more than it had in a thousand miles of inland waterways. The species were different, their proportions were different, they even occurred differently.*

The *Grampus* expedition solidified Ricketts's thinking about the relationships between animals and their environments and how, in marine settings, changes resulting from wave shock, bottom types, and varying tidal levels determine the ecological arrangements of the animals along the seashore. While this explains how he and Steinbeck studied marine life in *Sea of Cortez*, the genesis of this thinking came from Ricketts's time aboard the *Grampus* and it is most manifest in *Between Pacific Tides.* In addition to collecting a treasure trove of marine specimens, the *Grampus* expedition gave Ricketts enough evidence and information to write his now-famous "Wave Shock Essay." The differing shore topographies he discovered on the *Grampus* trip convinced him that there were three basic "co-ordinate and interlocking factors that determine the distribution of shore invertebrates:" (a) the degree of wave shock; (b) the type of bottom—whether rock, sand, mud or some combinations of these; and (c) tidal exposure. It is the impact of these "co-ordinate and interlocking factors" on life in the intertidal that defines the structural organization of *Between Pacific Tides.* The way Ricketts explains his new understanding to readers of this book is the reason that *Between Pacific Tides* remains a classic work of marine biology and an invaluable text for students and professionals in marine science, as well as for anyone who finds the shore a place of enchantment, wonder, and beauty.

There are few seashores in the world as beautiful as those along the Alaskan coastline, a fact not lost on the crew of the *Grampus* and, in fact, with everyone connected with the writing of *Between Pacific Tides.* This includes the principals, all of whom had connections with Alaska in or near Sitka.

Jack Calvin had a life-long love affair with Alaska that began when he homesteaded there in the 1920s; he relocated permanently from Monterey to Sitka in 1932. Ritch Lovejoy, who created the excellent drawings in *Between Pacific Tides,* lived in Alaska for a time and worked as a writer/illustrator for the *Arrowhead,* a Sitka bi-weekly newspaper. Jack and Ritch married daughters from the famous Alaskan Kashevaroff family. As we've noted, Sasha was a participating member of the *Grampus* crew and helped Ricketts with his collecting. Xenia, a third Kashevaroff daughter, was friendly with Ricketts, the Calvins, and Lovejoy. She and her husband, the soon-to-be famous modern musical experimentalist John Cage, helped Ricketts compile the index to *Between Pacific Tides*. In fact, the book has as many roots in Sitka and elsewhere in Alaska as it does on Cannery Row.

Sea of Cortez is an important book in many ways; it tells us a great deal about the littoral of the Gulf of California. Steinbeck's reissued version, *Log From the Sea of Cortez* with its important Appendix, "About Ed Ricketts," is a wonderful read—a combination of science and philosophy told with an arresting sense of humor that is rare in travel literature. The reality, though, is that however valuable the books are about Ricketts's [and Steinbeck's] contributions to marine ecology, they are a more polished restatement of the world view of *Between Pacific Tides.*

After the *Grampus* trip, Ricketts had only to assemble the data from that trip and integrate it with his already informed understanding of the central Pacific shoreline. Ricketts described all this in a letter to Torsten Gislén on October 21, 1932:

> *Our attempt, in which we seem to have been successful, has been to provide an authoritative and above all thoroly [sic] interesting account of the animals themselves, their life history and physiology, and their relation in communities to such environmental factors as wave shock, type of bottom,*

and tidal level. In addition to this field work, this has required considerable literature searching. Some 500 papers and books were examined; more than half were read, abstracted, and listed in the comprehensive bibliography.

Read just about any part of *Between Pacific Tides* that deals with animals who live exposed to rocky shores on the open ocean and you'll find striking similarities between Ricketts's observations about wave shock recorded during his trip on the *Grampus* with those in *Between Pacific Tides.* During and immediately following the *Grampus* expedition, Ricketts wrote his "Wave Shock Essay," "Notes and Observations, Mostly Ecological, resulting from Northern Pacific Collecting Trips Chiefly in Southeastern Alaska, with Special Reference to Wave Shock as a Factor in Littoral Ecology." Unpublished for the better part of a century, it was brought into print as a primary feature of the 2015 first edition of this very collection about Ed Ricketts. Of particular interest in Ricketts's essay is his thanks to Joe Campbell for "a clear and concise statement of the situation [that] all of us observed and discussed," namely that

the progressive assumption of supremacy in each habitat by the animals better fitted to withstand its peculiar limiting factors would progressively augment the survival difficulties which would have to be confronted by the others. ...The better fitted would be giving battle, furthermore, in a terrain favorable to themselves and peculiarly unfavorable to their enemies.

Considering this is Campbell's prose, we can see the still-novice mythologist integrating myth with science in language not dissimilar to Ricketts's statements about the terrains of battle among humans as well as starfish.

However important his collecting expeditions along the California coast and into Northern Mexico, as well as along the shorelines of the Pacific Northwest, there were no collecting sites more valuable than those he, the Calvins, and Joseph Campbell explored during their trip on the *Grampus*. The wide diversity of habitats they encountered provided Ricketts with the best possible sites for his field research. His conversations about the animals he collected with the Calvins and Campbell led to new insights into marine ecology and beyond even science itself to the boundaries of the metaphysical. Years later, after Campbell had become the world's most prominent comparative mythologist, he wrote that it was with and because of Ed Ricketts that he came to understand the integral relationship between biology and myth that, in part, defines his study of the archetypal hero in *Hero with a Thousand Faces* (1949).

Similar to his celebrated friendship with John Steinbeck—in which Ricketts and Steinbeck shared their love of words and of life—Ricketts's relationship with Campbell, which began in Carmel and bloomed full flower in Alaska, was symbiotic. Two men of acute intelligence, with the most inquiring of minds, learned from each other. That learning helped Ricketts (and Calvin) shape *Between Pacific Tides,* and led to Campbell's emergence as America's most revered comparative mythologist. However important the science Ricketts did along the coastlines of Vancouver Island and north to Sitka, his conversations with Joseph Campbell may be the most expansive achievement of the *Grampus* adventure.

There are a good many details we don't yet know about the time Campbell and Ricketts spent in Alaska and on the *Grampus*. Much can and will be gleaned from the Campbell materials in the New York Public Library, but these are beyond the scope of this brief study. We do know, though, that some years after his time with Ricketts in Alaska, Joseph Campbell wrote the draft of an unfinished novel he called "The *Grampus* Adventure"—an interesting title since, in his masterful *Hero with a Thousand Faces,* he not only refers to the hero quest as a journey, but also as an "adventure." We also know that a recurring theme throughout Campbell's work, the notion that "life lives on life," owes much to his observations in Ricketts's lab about the manner in which Ed fed mice to snakes. This notion was central to his fusion of biology and myth. Above all, we know that both men were insatiably curious about almost everything they saw and discussed, that exploring the curious broadened the horizons of both men. As an example, more than a decade after their time together in Alaska, when Ricketts read Campbell's *Skeleton Key to Finnegan's Wake,* he concluded that James Joyce's novel was the most important book of its time.

Also, it was during their time together that Ricketts was working on three philosophical essays that, taken together, define his view of life—the manner in which, as Steinbeck described, he looked from the tidepool to the stars and back to the tidepool again. One of those essays, the piece on non-teleological thinking that found its way into the *Log from the Sea of Cortez,* champions the value of "is" thinking—looking beyond what could be or should be to what is. Katie Rodger tells us that, "On the *Grampus*, he [Ricketts] shared his early notions of 'breaking through' and 'non-teleological thinking'" —philosophies that shaped his worldview based partly on his readings and understanding of Zen Buddhism and the Tao Teh Ching. Consider, then, Campbell's response to Bill Moyers during *The Power of Myth* PBS television interviews when Moyers asserts to Campbell, "You changed the definition of a myth from the search for meaning to the experience of meaning."

> *Experience of life. The mind has to do with meaning. What's the meaning of a flower? There's a Zen story about a sermon of the Buddha in which he simply lifted a flower. There was only one man who gave him a sign with his eyes that he understood what was said. Now, the Buddha himself is called "the one thus come." There's no meaning. What's the meaning of the universe? What's the meaning of a flea? It's just there. That's it. And your own meaning is that you're there. We're so engaged in doing things to achieve purposes of outer value that we forget that the inner value, the rapture that is associated with being alive, is what it's all about.*

For Campbell as for Ricketts—and for John Steinbeck, largely via Ricketts—living was about participation, the need to live life deeply. To, as Campbell often told his students at Sarah Lawrence, "follow your bliss." What makes the lives of both Ed Ricketts and Joseph Campbell so special—and defines the Ricketts persona as well as Campbell's singular exploration into the power of myth—is the fact that both men did what Campbell told his students: they "followed their bliss." And all this—the genius of Ed Ricketts and Joseph Campbell that emerged as they explored the rocky coasts of Southeast Alaska, and as Ricketts conducted his important fieldwork—is what defines the particular brilliance of Ricketts and Calvin's *Between Pacific Tides* and Campbell's *Hero with a Thousand Faces.*

Notes and Observations, Mostly Ecological, Resulting from Northern Pacific Collecting Trips, Chiefly in Southeastern Alaska, with Special Reference to WAVE SHOCK as a Factor in Littoral Ecology

E. F. Ricketts
Pacific Biological Laboratories, Pacific Grove, California

The current Pacific Biological Laboratories on the waterfront in Monterey, rebuilt after a fire in 1936 destroyed the original. *(Courtesy of Nancy Ricketts.)*

ABSTRACT

A short account is given of the littoral biology of the inland passage to southeastern Alaska, with especial attention to differences between the faunas of the inlets and of the outer shores, as exemplified particularly by Sitka. The outstanding invertebrate animals are mentioned by reference to occurrence, abundance and natural history. There is some specific ecological information with reference chiefly to coelenterates and echinoderms. Some natural history traits, and possible indices of occurrence are stated for *Aurelía* and *Gonionemus*; formae of *bunodid* [*bunodactis*] anemones are considered.

An ecological classification is suggested, based on the occurrence and intensity of wave shock. The importance of the wave impact factor is stressed for the distribution of shore invertebrates, culminating in a generalized statement thought to reflect the environmental natural selection situation with reference to agitated and quiet water communities. The interrelated nature of the separates comprising the shore physiography-type of bottom-current-wave shock complex is mentioned.

I.

For some time, the writer has been making annual collecting and observing trips into the Puget Sound-British Columbia region, culminating this summer (1932) in a ten weeks trip, mostly by small boat, into Southeastern Alaska.[1]

In addition to some commercial collecting at Sitka, the desiderata were: to make a hurried ecological reconnaissance of the shore invertebrates, especially with regard to such environmental factors as wave shock, type of bottom and intertidal zoning; and to procure representatives, for future identification when necessary, of the ecological "horizon markers", and of taxonomic groups now being actively studied by specialists. It seems to me important for collectors to devote more than usual attention toward procuring Pacific sponges, polyclads, amphipods, hydrocorals, sipunculids, rhizocephalans, etc., since it may be several generations before systematists again focus on these groups, [socially] known (among many others for which there are no champions) here on the Pacific coast. There seemed also a chance to have the *Bunodactis-Cribrina-Evactis* situation straightened out; hence we paid particular attention to these actinians.

[1] With the exception of a few alterations to punctuation for ease of reading, the "Wave Shock Essay" appears here exactly as written by Ed Ricketts. This includes some outdated spelling of place names, etc. and the author's unvarnished observations of Alaska Native life. While it is possible that changes might have been effected by editors if Ricketts had published the essay during his lifetime, we have chosen to present it as it was left by him over eighty years ago. There were, however, a few words that could not be clearly identified during the transcription process and in those instances, brackets are present in the text to denote the uncertainty.

The British Columbia-S.E. Alaska shore is interesting biologically in that it has received scant attention recently, although it was one of the first areas on the Pacific Coast to be carefully examined (Brandt[2], the Russian, in 1835). In the groups treated, the reports of the Columbia Puget Sound Expedition, the Harriman Alaska Expedition and those of the Canadian Arctic Expedition are illuminating, and there are various other scattered papers (Verrill[3], etc.) and from the marine stations at Friday Harbor and Nanaimo. Thus a good many systematic surveys have been made and reported on by specialists; but seldom in a general way or ecologically—the point of departure having been specific rather than general.

The mid-area is incidentally interesting for its lonely and therefore hospitable settlers, a group which must be spotted earlier on the culture curve than the megalopolitans with whom most of us are in contact. S.E. Alaska is more densely populated, and has several large towns, but in central and northern B.C. it is possible for one sympathetically to observe his immediate forebears, in the actually pioneering conditions of 50 years ago. There are no roads, and of course no railroads (excepting at Prince Rupert); civilization here is still in the packtrail- steamboat era; and of the modern appliances only the radio has taken hold. All this in spite of the tourist steamers that serve Prince Rupert and S.E. Alaska through the inside passage; the main steamer routes only have been adequately charted—the land almost not at all. Dotted lines on the charts mark the approximate position of many small rivers and inland lakes; small islands and minor passages. The fjords and precipitate mountains are

[2] Publisher's Note: Johann Friedrich von Bandt
[3] Publisher's Note: Addison Emery Verrill

scenically very noticeable whenever the rains lift sufficiently to permit visibility. Finally, the Indians are apparently well adjusted to their changed environment (necessarily affected mechanically and sociologically by contact with whites). By contrast with Indian Bureau scandals in the states, they are sensibly treated (especially in B.C.), and noticeably so by the missionaries, at least many of whom are tolerant and who foster their truly worthwhile handiwork. One suspects that any careful observer in out-of-the-way places here will come away with changed ideas regarding missionaries, Indians and officials.

Grateful acknowledgements are due to a number of interested and interesting individuals along the way, by whose aid a fine cruise was made finer; foremost of course, to Jack and Sasha Calvin, skipper and mate of the dry decked (but not really too dry) good and congenial motor launch *Grampus*; to Father Kashevaroff, Curator of the Alaska Historical Museum for really exceptional hospitality and assistance at Juneau; to the Thompson family and especially to Kit Thompson for a pleasant stay and interesting collecting at Refuge Cove; to the Bolshanins at Sitka, and to George Sumption, master of a shrimp netter at Wrangell. I have especially to thank Joseph Campbell of New York, a constant and interested companion, for clarifying and outlining in some detail certain conclusions attained en route. For more recent cooperation, I am grateful to Dr. W.K. Fisher, Director of the Hopkins Marine Station, who was good enough to read over the manuscript carefully and to make several significant suggestions, to G.E. MacGinitie, director of the Kerckhoff Marine Station, and to Dr. Waldo Schmitt of the United States National Museum for similar kind offices.

II.

Starting in the latter part of June from Tacoma, we cruised to Juneau by easy stages, traversing the inland channels via Nanaimo, Pender Harbor, Alert Bay, Prince Rupert, Ketchikan, Wrangell, Petersburgh and Sitka.

Topographically (and hence to some extent biologically) the area covered is divisible as follows:

A. Puget Sound and Lower B.C. to Queen Charlotte Sound
B. Middle and Upper B.C., to the Skeena River near Prince Rupert
C. Southeastern Alaska, (1) inner and (2) outer coast

A.

From Puget Sound to the crossing at Queen Charlotte there is an increasing erosion gradient.

At Tacoma there are chiefly depositing areas, shoal quiet water bays with mud bottoms and gradually sloping shores. The occasional reefs have the *Cucumaria-Ophiopholis* and *Terebratalia-Serpula* communities which are so well developed at Friday Harbor and northward. Soft bottom facies are commonest; with the crab *Hemigrapsus oregonensis* (and *nudus*), the butter clam *Saxidomus*, the burrowing shrimps *Upogebia* and *Callianassa*, and the worms *Glycera*, *Pectinaria* and others in the mud; the sand dollar *Dendraster* and the moon snail *Polynices* etc. on the sand; *Balanus glandula*, *Acmaea cassis pelta olympica*, crusted *Mytilus edulis*, Nereis, the rock cockle *Paphia* in the substratum, *H. nudus* (and *oregonensis*), and, submerged, *Evasterias* and the cucumber *Stichopus*, on the gravel. Of pelagic forms, thick *Aurelia*, *Aequorea*, *Phialidium* and *Pleurobrachia* are variously and locally common, often in restricted and delimited communities in a distribution difficult of interpretation. One bay will harbor mostly *Aequorea* at a given time, another, apparently similar, only *Mitrocoma*, the biological (rather than physical) factor seeming the most important.

The San Juan-Straights of Georgia region has both depositing and eroding shores[4]; the fauna there has been described frequently. On the southern shore of the Straits of Fuca previous collecting trips had indicated the presence of an excellent area west of Port Townsend, where thousands of the sessile schyphozoan *Haliclystus* are a feature of the littoral algae, and again at Pysht, not far from Cape Flattery but well inside the entrance to the straits. Here the bouldery beach is very wide, certainly half a mile or more, with a gradual slope, and certain whole assemblages of animals are reminiscent of the 70-80 fathom bottom at Monterey, when some great rock, its interstices filled with silt, is brought to the surface. There is complete protection from wave shock, but the currents are good and the water cold and oceanic. At Clam Bay, south of Nanaimo, July 2, the tectibranch *Haminoea vesicula* just about at the extreme north of its range, was very much a feature of the eel grass beds.

[4] In this account the word "shore" is taken to mean the intertidal—or littoral in an ecological sense. It happens frequently that the foreshore exposed at low tide will be entirely rocky, while below that, gravel or mud shore will slope away gradually to the bottom; such a region would be described as "rocky shore."

There must have been tens of thousands of these bubble shells, found also in similar situations in estuaries along the California coast.

Single individuals of the large and showy hydromedusae *Polyorchis* were picked up occasionally along these waterways.

North from Pender Harbor (and I understand Campbell River on the west wide of the Strait) to Queen Charlotte Sound the shores are mainly rocky, hence with eroding shore communities decidedly in the ascendancy. Culminating locally in Agamemnon Channel (49° 42' N, 124° 5' W), we turned up a very rich boulder fauna featuring *Evasterias*, the brachiopod *Terebratalia transversa* in great numbers, the leather star *Dermasterias*, the rock oyster *Pododesmis*, *Crepidula nummaria*, *Bugula murrayama* [?], the stalked tunicate *Styela gibbsii*, many chitons, the channeled encrusting yellow sponge *Mycale*, and occasional jointed bryozoa probably *Cellaria* such as we dredge in Monterey Bay. Where the reefs are solid, or where vertical walls deflect the currents, the common seastar *Pisaster*, twisted colonies of the tube worm *Serpula*, and the vividly red tunicate *Styela stimpsoni*, with the usual *Mytilus*, and medium to giant *Balanus cariosus*, seem to be the most characteristic. Part of this region, in the very quiet water rocky coves (i.e. Refuge Cove) was interesting subtidally in that quantities of an erect and often branching vividly coral colored sponge (*Esperiopsis* or similar) were taken at low tide via herring rake. Occasional specimens reached a height of 10". A forest of these in almost pure culture seen through clear water, is a lovely sight more reminiscent of the tropics than of these cold waters. A land locked sea water lake at Squirrel Cove, so completely cut off that the tide must come in and out through a narrow channel and with its low tide level several feet above the surrounding sea level, had a most remarkable concentrated fauna of *Stichopus*, *Dermasterias*, and enormous *Terebratalia*, much darker in color than usual, with the small green shrimp *Spirontocaris paludicola* abundant. The confining channels of the Ucultaws and other rapids are plastered with vividly violet *Pisaster*, and the shallow bottoms in the swift parts have *Strongylocentrotus franciscanus*, called "sea eggs" locally. These very rapid water areas (with current velocities occasionally up to 10 knots)—Seymoure and [??]taws with their [????] ten in the air at once—could easily provide US migration barriers to such delicate attached animals as that, gravel or mud shore will slope away gradually to the bottom; such a region would lack pelagic larvae, but we found no specific evidence of this. In navigating a small power boat through such rapids, one cannot fail to be impressed with the terror that must have plagued the early navigators, in sail or row boats, suddenly swept into these swift currents, with no knowledge of slack water periods, no method of controlling their craft, no way of knowing what submerged rocks might be ahead, or how long before the shores would widen out again with quiet water. At Port Harvey, south of Alert Bay, we found an area in which *Aurelia* was apparently endemic. The distribution of this probably gregarious form is considered in Part III.

B.

Middle and upper British Columbia, from Queen Charlotte Sound to the Skeena River represents the acme of ruggedness, and an almost total lack of depositing shores. The channels are deep, straight and comparatively narrow;

the shores are precipitous, sloping imperceptibly into mountains that may attain almost a mile elevation within three miles of the water. There is no beach; the intertidal zone is narrow and differs topographically not at all from the shores above and below, so that frequently a fairly large boat can be moored for the night to a tree, if unable to anchor because of depth. Talus falls directly to the usually level channel bottom which achieves great depths (nearly 50 fathoms, often to more than 100) only a short way from shore. Currents are good, sometimes very swift. Rainfall is heavy, 80" certainly, and probably over a hundred in places. The *Canadian Pilot Book* records 2054 mm for Prince Rupert, with an average of 221 rainy days per year. There are many waterfalls, not so good for shore animals, but carrying lots of organic detritus for bottom living forms. It would be easy to predict good dredging here. Along these channels, to the great discomfiture of deck hands intimately described here as connected with the galley stove, no driftwood was available, whereas it had been everywhere south of Queen Charlotte Sound.

Scenically there is a striking resemblance between this region and parts of Yosemite; the same rugged slopes, the same appearance of heavy foresting by conifers; all the very steep slopes and exposed rocks are of granite; there are many waterfalls and snow-covered peaks. And on such rare occasions as the air is clear of rain, it can be brilliantly clear.

At Bella Bella Cannery, Indian Village, and store, where at least one old wood carver is doing excellent work on cedar chests, and where the morale of the Indians is high, if one may judge by their clean and interesting "lived-in" houses, there is a bit of low and rolling country, and some depositing shore.

On July 14th, specimens of the preposterous pelagic nudibranch *Melibe leonina* were taken here in an eel grass cove. These are large (to 5 or 6"), fragile, transparent and delicate, but they swim well if grotesquely by snapping the body sideways. The hood, filled with air, is said to serve as a float, but the animal can also attach to such support as eel grass with some force, wrapping its foot around the stalk

Elsewhere, so far as known, the soft bottom shore formations of strictly marine animals are few and far between in this precipitous region. Depositing shores occur at the heads of inlets, but here the sand or gravel is continually washed by floods from fresh water streams, especially at low tide. The clams, anemones, etc., that have been able to colonize these areas are periodically deluged with fresh water, and always live pretty close to their fresh water lethal (or at least inactivity) point.

At Fishermans Cove (53° 20' N, 128° 50' W), north of, and on the shore opposite Butedale Cannery, we found the richest bouldery shore collecting thus far noted. There were *Cucumaria miniata* literally by the thousand—the test specimens fairly easy to narcotize—with two or three *Ophiopolis aculeata* for every holothurian, and often literally clinging to them. Nemerteans were abundant. There were hundreds of the sipunculid *Physcosoma*, larger however than those we take on the California coast although apparently of the same species according to Dr. Fisher, *agassizi Keferstein*. A giant flatworm, brilliantly brown, possibly *Cryptophallus magnus* Freeman, sometimes 3 or 4" long, was more than occasional. Directly under the rocks were large tubed terebellids, often with commensal polynoids and pea crabs (*Pinnixa tubicola*).

Paphia occurred in the substratum. Shelled snails were especially abundant; *Thais lamellosa* and *Searlesia dira* both by tens of thousands, lower in the littoral *Purpura foliata*, and submerged except at lowest ebb, the hairy *Argobuccinum oregonense*. *Katherina tunicata* occurred here as an anomaly, it's certainly found typically in its greatest abundance as an open shore form, but in this quiet water it was the commonest chiton. *Evasterias* was a common asteroid, but farther along on the bluff was *Pisaster*. The sandy flat at the mouth of a little creek had *Saxidomus*, or *Cardium* where pebbles occurred. The "*artemisia*" form of *Bunodactis* was frequently found attached elongatedly to the larger buried stones.

The attenuated dawns and twilights, the continued drizzly rain, and the thrushes singing for hours at night and morning from the wet and steep hillsides—the only sound in this quiet region, aside from the rush of waterfalls—are the things I remember chiefly from this country. Specifically at Lowe Inlet, a nice illustration was provided of the fresh-water limiting factor.

Located in a sub-channel off the long and narrow Grenville Channel, and with a booming waterfall at its source, the most seaward portion must be upwards of 30 nautical miles from true oceanic water, with narrow entrances and passages between, many streams and much rain. The following quotation from the July 19th collecting report is illuminating:

> *It would be easy to believe that fresh water is the chief limiting marine factor in these inlets. It was noted here with* Urticina, *and at Fishermans' Cove with the* artemisia *anemone and with cucumbers, that the water they*

were placed in for anaesthetization seemed to be injurious. They wouldn't expand. And they got the look of marine animals placed in fresh water. I tasted the water and found it nearly fresh. Of course this surface water is the freshest of all; it would naturally be the stuff picked up by bucket for narcotization trays, and it would be the stuff that covers the lowest littoral at the low tide. At high tide, these animals are under 15 or 20' of water, and hence in a medium of greater salinity than any we can reach. At low tide, I noted that practically all the submerged Urticina *near the surface were tightly contracted; a few above the water line were hanging limp and half open, in the way that anemones normally do when exposed to the air. It may be that many of these animals, having penetrated the inlets as far as they were able, are pretty close, at low water, to their lethal fresh water dilution point, and go through a period of necessary quiescence at ebb. However, they narcotize very readily when they do expand, (and we readily got them to expand nicely by immersing them in water from nearer the ocean); and I suspect that this is due to the fact that easy conditions especially in the lack of wave shock, allow them to become less "hard boiled" than they are on the open coast, for instance at Monterey.*

It has certainly been substantiated, on this trip and previously, that all these quiet water animals are easy to preserve expanded, as compared to open coast individuals of the same species. The Squirrel Cove *Stichopus* "went under" with a little Epsom salts within a few hours, whereas it takes *Stichopus* 24 hours or more at Laguna Beach, with uncertain results, and much longer at Monterey, with results however practically certain (the wrong way!). Some of the Sitka open coast animals were likewise difficult to work with, and the

Cucumaria there, even from fairly sheltered waters, were more refractory than those from the inlet at Fisherman's Cove. It seems to be the case that quiet water animals are "hard boiled" to fresh water, stagnation and detritus, but amenable to narcotics; whereas the shore forms fronting the open ocean resist nicely all chemical attempts to kill them expanded, but die more or less contracted, albeit quickly, in fresh or deoxygenated water, although sensitive to these factors.

In all regions so far considered, the absence of wave shock is an important factor. There is no really heavy wave impact in any of these inland waterways, even where there are straits and wide channels. Tenderly constituted yachtsmen who have crossed some of these stretches under difficult conditions will no doubt deny this allegation more spiritedly now, on dry land, than when the wave shock impressions were being formulated. Queen Charlotte and Milbanke Sounds, and Dixon Entrance, connecting directly with open ocean, do of course exhibit true wave shock through the ground swell—the sole source of surf—but this never occurs in inlets. No matter how the inland waters may be riled by wind, current or tide rip, it should be noted that there is never much white water inshore, nor any pounding sufficient to detach or crush even fragile sessile forms. Many of the animals characteristic of surf-swept coasts are entirely lacking here. Their place is taken by forms which may not have specialized in wave shock protective adaptations (unnecessary here) but which animals are nevertheless highly competent in other ways apropos to the region, ways which the surf living forms will have been able to neglect with no untoward results in their particular terrain and against their particular competitors. This whole situation puts an emphasis on the ecology of inlets that will be considered later.

C.

Topographically, the southeastern Alaska coastline, not so severe as the region just considered, reminds one of the San Juan Island region in Puget Sound. There may be here also many small flat islands; this is especially noticeable en route to Sitka from the north and east; there must be literally thousands of them. From the Skeena at least to the Stikine River, the shore is noticeably gentle, especially by contrast to the bold and barren regions just to the south. Environmental factors are otherwise the same, except that tides diminish slightly (spring range 17.9' at Prince Rupert, 15.4 at Ketchikan, 16.2 each at Wrangell and Juneau), and rainfall, if possible, increases (179" at Ketchikan). This far north (N 58° 18' at Juneau) the amount of daylight during the summer is considerable, but since natural sunshine is noticeable chiefly in the lack, this environmental factor should be of slight importance to littoral communities. Locally, floating ice from "live" glaciers (these actively discharging bergs into the sea water) undoubtedly makes a surface water hazard both through temperature and fresh water factors.

Canoe Pass, some 30 miles north of Prince Rupert, had an interesting pelagic fauna.

Aequorea was extraordinarily abundant, one or two *Polyorchis* were seen, and a few *Gonionemus*, plus the usually smaller medusae and ctenophores. This narrow passage however was chiefly spectacular in its display of *Cucumaria miniata*. We had here the opportunity of remarking its strict bathymetrical distribution, cut off sharply below at -1.0' and above at +3.0'

(Canadian standard; their datum is 1.9' below ours). This cucumber, seen in great beds just below the surface of the water, is quite a different animal than one would suspect from seeing it exposed at low tide, or from seeing preserved specimens, however well expanded. The tentacles are translucent, brilliant coral red or purple, they can be very completely and diagrammatically extended to form more or less a stationary net. These out-stretched tentacles are apparently adapted to the capture of delicate and mostly living minute pelagic organisms, rather than for sweeping the floor for detritus as in *Stichopus*. This probably accounts for *Cucumaria's* strict vertical limits below, a function of its food-getting mechanism, since the active and small macroplankton on which it feeds, live, in turn, on the microplanktonic diatoms, etc., produced only in the upper 28 or 30' layer of the water through photo-synthesis. (Shelford and Towler, 1925, Publ. Pug. Sd. Biol. Sta., p. 12, report the *C. miniata* association as extending down to 50 M, at Friday Harbor and many individuals unquestionably do reach that depth, but we found all the rich beds, wherever we investigated this form, cutting off sharply a little below the extreme low tide mark.) Hence, the lower limit is thought to have nothing to do with a need for the tidal rhythm; it is due to the workings of a biological (food) factor. The upper limit, however, is probably a function of increasing tidal exposure. Another type of association, similarly zoned, might theoretically be restricted to these depths by its sole dependence on algae for food—the red abalones represent such an association along the California coast—but if it occurs here at all it must be limited to such forms as the red sea urchin on agitated shores. Below these two lowest littoral associations and in their lower reaches competing for placement, is another fauna presumably less rich (and often actually so, if low tide observations through clear water are indicative), peopled partly by animals which depend on the rain of mostly dead material washed in and dropped down, partly by the starfish—brittle star—*Stichopus*-annelid detritus feeders, which pass great quantities of the bottom muck through their guts, extracting the contained nourishment, and partly, possibly, by actively predaceous forms which feed on the others.

Because a small shrimp trawler was working at Wrangell, we stopped for several days at this unspoiled, not self-conscious town. But in the curio shops "Souvenirs of Wrangell" are little packets containing Panamanian shells! Many California abalone shells had similar captions. The Lord preserve the concepts of tourists, or whoever buy such things! Fifty cent and one dollar totems made in Japan to sell, (the Indians can't work so cheaply). Some very fine local ones, at a price.

I heard one small Indian boy remark to another (there are two adjacent churches): "These Sunday schools look like they're both the same." Their theology must be naive, and amusing as too much of ours is not. I think some of the Indian girls are beautiful; certainly with an atmosphere that to us means inscrutability; they walk like Pallas Athene. It has often been stated that interracial marriages—not liaisons, but permanent relations—are common here, with few break-ups or even infidelities. I have come to regard this situation, if true, not (necessarily) as a stop gap on the part of a lonesome white man, but actually as his weighed desideratum. If successfully transplanted, he becomes part of the country, sees it through wilderness rather than Seattle or London eyes, and sees the Indians as fittingly part of that country which he likes and into which he is expandingly fitting. His native wife is cherished because she gives him in richness of living, and in (probably unconscious) spiritual companionship, more than the women to whom he is collectively accustomed, compensating thus her cultural failings, so that the relation is complete and desiredly unbroken.

The dredged fauna here is noteworthy. This is possibly the only place in America where quantities of living Rhizocephalans can be taken consistently; *Peltogaster* and *Sacculina* commonly on crabs, *Sylon* on shrimps more rarely. Clusters of *Allopora* several feet in diameter are frequently taken, tho shrimpers however avoid known areas of these branched and vividly red hydrocorals with a view to conserving their nets. There are some pennatulids (the big *Pavonaria* with its commensal actinian encircling the stem) and coral red gorgonians. *Gorgonocephalus* is exceedingly abundant; *Ophiura* of course by the thousand, some *Stichopus* and *Strongylocentrotus drobachiensis*, the last larger than any littoral specimens I have seen. The giant scallop *Pecten caurinus*, with shell as large as a dinner plate, may be locally prevalent, the short squid *Rossia* (called cuttlefish and devilfish locally), *Cancer magister*, etc., *Pendalus borealis*, the pink shrimp, *P. hypsinotus*, the rose or hump shrimp, and *Pandalopsis dispar*, the side stripe, were taken in that order, the first predominating, along with many other *Pendalus*, *Crago*, and *Spirontocaris*. The bastard shrimp of the fishermen turned out to be large *Spirontocaris brevitostris*, and the highly ornamented rock shrimp to be *S. groenlandica*. The fishermen inquired of me why it was that they failed ever to find small females or large males of a given species, and I couldn't then provide the answer, recently determined by A.A. Berkeley (Contrib. Canad. Biol. & Fish.,

NS, VI, No. 6, 1930) at the Pacific Biological Station, Nanaimo, that sex reversal takes place at from 26 to 42 months—depending on the species—from male to female, in each of the five species studied.

Petersburgh, another center of shrimp trawling of which we were unfortunately unable to take advantage, exhibits a striking faunal change in the littoral, due, one suspects, to the exceedingly cold and brackish water incident to melting ice. The water "looks" different, it severely chilled our hands during a short reconnaissance of the float fauna; even the avifauna is noticeably different. There were now hydroids and bryozoa on the float, even the anemones were different in appearance.

The Juneau fauna had been touched in the examination of previous catches, so no stop was made here on the way up. At Sitka—the finest collecting place I have ever seen—the semi-protected coast waters turned out a rocky shore assemblage with inlet components, but mostly derived from the coves of the outer coast, [finding?] *Haliotis kamschatkana* infrequently, and *Strongylocentrotus franciscanus* not at all; *C. miniata*, *Ophiopholis*, giant flatworms, nemerteans, *Physcosoma*, *Evasterias*, *Pisaster* (one or the other markedly dominant locally), *Thais lamellosa*, etc. On flats with rocks, the apodous holothurian *(Hiridota albatrossi)*, the sipunculid *Physcosoma agassizii*, the probascis worm *Glycera*, the permanent burrows of *Echiurus pallassii*. (At 11 PM, Aug. 3, 1932, we found a small (2") *Echiurus*, free swimming, writhing and contorting in the water, at Jamestown Bay near Sitka.) with *Pinnixa schmitti*, the small red, smooth and gregarious anemone *Charisea*? *Strongylocentrotus drobachiensis*, nemerteans, *Evasterias*. The more gently sloping flats of true mud, stalking ground of the lonely heron, had great beds of *Echiurus* and *Arenicola*, *Glycera*, pelecypods such as *Saxidomus*, the moon shell *Polinices*, etc., as at Jamestown bay. Eel grass beds had *Gonionemus*, at least four shrimps—*Spirontocaris camtschaticos, Sp. Paludicala, Sp. cristala* and the attenuated *Hippolyte californiensis, Obelia,* snails and hermits, eel grass limpets, *Pentidotea*; *Stichopus* and *Mediaster* subtidally. In the coarse "corn meal" sand: *Cardium*, sand dollars and the elongated "*artemisia*" fern of *Bunodactis* attached to deeply buried round stones. A summary of the fauna of "inside" waters would stress the scarcity, as compared to the California coast for instance, of crabs, especially of the high foreshore inhabitants, of hydroids (other than *Obelia longissima* and *Gonothyraea* on floats), and of anemones (other than the *Metridium* of piles and floats) especially by comparison with the great beds as at Monterey. Many of the richest collecting grounds are on floats. These must be important factors in dispersal through a medium unfavorable otherwise. In the dissemination of pelagic larvae(of competent introduced forms, for instance) a few take roots on the floats (finding thus favorable attachment sites in a region otherwise unadapted due to the small amount of foreshore), survive, and as adults produce larvae which likewise radiate from the new center, thus eventually searching out and populating all favorable areas more rapidly than would be possible in the absence of floats.

C-2

The rich pelagic associations fade as one approaches open water (on the one hand; and fresh water on the other, for that matter), and in the channels approaching Sitka, oceanic jelly fish predominate, *Phacellophora* or similar.

In its strict delimited fauna (as regards wave shock) Sitka provides a really excellent text book illustration for ecology. A short trip of 3 of 4 miles from the Sitka inner harbor to the fringing Kyack Islands or to Whale Island puts one into a new and more brilliant littoral world. Even where the substratum is similar, there is an utterly different set of animals.

In its features the surf swept outside fauna is almost identical with that of our Pt. Lobos to Pt. Sur California region, except for the addition of *Gersemia rubiformis*—the red encrusting alcyonarian, and certain [renk?] large plumularians. There are many sponges, especially *Reneira cineria* and the red sponges, *Abietinaria, Eudendrium* and other hydroids, *Stylantheca porphyra* Fisher (hydrocoral), *bunodactis xanthogrammica*, a few *Charisea, Tagula*, [? ?], *Flustra* and other encrusting forms among the bryozoa, *Pisaster*, giant *Balanus* almost replacing the *Mitella* so common further south but which also occur sparingly here, *Str. Drobachiensis*, an occasional *S. franciscanus*, chitons especially *Katherina, Mytilus californicanus, Thais emarginata, Calliastoma costatum*, nudibranchs especially *Diaulula* or similar, compound tunicates, not many crabs but a large percentage of those examined had *Sacculina* or *Peltogaster*.

On the inner (lee) side of Kyack island, there were different hydroids and bryozoa, especially the daintier forms, other sponges, but *Reneira* VVP [very very prevalent], *Acmaea, Charisea*? and the *elegantissima* form of *Bunodactis* (as contrasted to the *xanthogrammica* form on the outside) and many less obvious animals. An "outside" cove, well protected from wave shock by two narrow entrances, as at Cape Burunof, 14 Aug., had no anemones other than *Urticina*, (a few *Polyorchis* were seen here also), no alcyonaria, plentiful

Gonionemus clinging to the stems and stipes of a scraggly and large brown alga, bryozoa as above, the ubiquitous *Physcosoma*, a tube building Nereid, terebellids with commensal crabs *Pinnixa tubicola*, giant *Pycnopodia* and *S. franciscanus* VVP., many *Pisaster* but only one *Evasterias*, crabs abundant but especially *Lophopanopeus bellus*, and including a *Cryptolithodes* and *hapalogaster mertensii*, chitons including *Cryptochiton*, keyhole limpets, *Haliotis kamtschatkana* (plentiful enough so that we took 40 or more, the largest a bit over 3" in length of shell, enough food for all of us.), *Thais lamellosa* and *Purpura foliata*; an exceedingly rich and varied assemblage, probably worth the scarred and itching faces we took away as a result of an encounter with [a] host of black flies that cloudy and quiet morning. These particular pests (a warning that will not be amiss to similarly situated voyagers) work quietly, almost painlessly, so that one is first aware of an attack by discovering a spot of blood where a good sized chunk of flesh has been removed. The animals flock in such hordes that it's impossible to avoid being actually mutilated if one is out unprotected when and where they occur. The aftermath is delayed for hours to overnight, when a really considerable itching sets in, secondary infection is likely, and one carries bruises for weeks, and scars for months. The situation is serious enough to justify one's packing constantly, and using whenever the insects are flying, any heavy and odorous tarry ointment that will discourage the less persistent of the beasties—some are sure to get through anyway.

III.

In attempting to formulate some idea as to optimum localities for given common species, again on this trip I was impressed with the necessity of one competent person covering the region pretty thoroughly before attempting to evaluate the comparative distribution of even a single significant animal. Literature records of range are often inadequate; even those by competent fieldmen who consider abundance. The red starfish *Patiria miniata* is listed for instance by Johnson and Snook (p. 206, Seashore Animals of the Pacific Coast, Macmillan, NY, 1927) from Sitka to Lower California, no data as to depth, presumably littoral. Fisher records it (p. 257, USNM Bulletin, 76, Pt. 1, 1911) from Sitka to above La Paz, low tide from 165 fms. Authentic specimens of course are known from both extremes, and this great range cannot be questioned, but the effective range is probably more limited. An observer who had collected only at Sitka, Puget Sound, northern California and (say) San Simeon Bay in Central California would likely turn out up to a dozen typical and large specimens per tide at each place, and he would be justified in rating this animal as abundant and widely distributed. A different observer equally oriented and competent, working at La Jolla, Ensenada and Boca de la Playa in Lower California would find in turning over rocks possibly half a dozen small specimens (which the northern man, could he see them, would know instantly for stunted and inhibited forms), and maybe one or two scraggly mediums per day. So he also would rate this as a relatively calm or under rocks inhabitant. The Monterey Bay worker, on one good minus tide, could find more individuals than both the others had in their seven hauls. Yet each independent collector would rate them common, and the bibliographic compiler of records would have a heterogeny from which to draw his generalizations. With regard to this particular group of animals, there happen to be adequate supplementary records, Fisher summing up the ecological situation in his remarks on p. 257. But in many groups there are only bare listings, with no way of determining even whether the species can be expected littorally or through dredging, and one can build unchecked whatever zoogeographical generalizations he inclines toward, the scant and undifferentiated facts will substantiate any generalized theory. It is manifestly true, as per Allee's 1923 contention (p. 189. Biol. Bull. XLIV No. 4, 1923) that if the search is long enough and competent enough almost any animal can be found in almost any environment, however anomalous; there are always the "strays." *Patiria*, specifically, seems to have an "effective" range primarily in the low littoral and upper sublittoral, from Morrow Bay northward, especially along the outer coast clear to Sitka in decreasing abundance, with the optimum around Monterey (but probably spotted elsewhere in favorable situations). Specimens occurring to the south of this effective range are few, mostly stunted and thoroughly sheltered under rocks or dredged—the untoward conditions obviously limiting their size, abundance and activities.

Ecological data of some value however are thought to have been secured, chiefly as a result of these northern observations, on the few following forms.

*Aequorea acquorea (Forskal)*Although recorded (in two distinct varieties, however) in the open ocean from such extremes as San Diego and Dutch Harbor (p. 39, 40, Bigelow, Publ. No. 1946—From Proc. USNM, 1913) this noticeable and vividly phosphorescent form seems to achieve its optimum abundance in the quiet inland waters along Puget Sound and the Straits of Georgia. It rarely occurs abundantly with *Aurelia*, either mainly one or the other occurring in a given small bay, and I have been unable to determine any obvious environmental character that causes the anomalous distribution. *Aequorea* however is probably the commoner of the two, and certainly the more ubiquitous, although possibly not so gregarious. It seems also to alternate with Aurelia in abundance from year to year. Although always present, it was particularly abundant in the summer of 1933, when not a single *Aurelia* could be found in their usual haunts in southern and [...]

[*Original essay is missing pages 25 and 26. Text resumes on page 27*]

[...] sprinkling of minute forms, by wading around in the deep eel grass beds in Puget Sound at tides of –1.0 or lower. Whereas the Atlantic species is reputed to swim to and from the surface throughout dull days, fishing small living macroplankton with its outstretched nettle celled tentacles, the Puget Sound form seems never to swim to the surface unless the water or the eel grass, especially at the slightly mucky bottom, has been just previously agitated. Otherwise they occur so far down that they risk no exposure at low tide, even on the –3.0 ebbs. In October- November 1931, not a single specimen was to be found, although search of the identical beds was intense and prolonged. The presumption is that *Gonionemus* has a summer and perennial life, dying out, as do so many of the Puget Sound forms, early in fall when the planktonic production suddenly falls off.

Wherever we have made careful search in the beds of deep eel grass in suitable regions we have found a few, but quantities seem to be present only in localized little pockets, earmarks of which are as yet unrecognized. In 1932, one or two sexually mature specimens each were taken at Clam Bay (South of

Nanaimo) and at Refuge Cove, both in early July. At Squirrel Cove, it was reported that literally hundreds had been seen lying on the eel grass and kelp at low tide, in the rain, early in June. The few specimens we were able to take in July were hard come by; we agitated the bottom algae at low tide with a long handled rake, capturing the detached animals with long handled nets as they swam away or to the surface. Later on, further north, we searched through several eel grass beds in the rain (it's surprisingly easy to select a rainy day up here!), but entirely without success. A few more were taken at Canoe Pass, above Prince Rupert. They were seen again at Sitka, and here they occurred in greater quantity than in Puget Sound, both inside and on the fringing islands. Those taken at Jamestown Bay, Sitka, early in August, were mostly smaller than usual, but sexually mature. Many were taken from eel grass beds, but the bulk of the animals were captured by agitating violently with a long handled net, the fronds and stipes of a wide and veined dirty brown alga (Alaria sp?) growing from –3.0 to –10.0', and hence available only on very low tides. The conditions of collecting, mostly during the night by spotlight, and in rain or wind, were such that it was possible to see the animals only after they were detached and swimming. However, on the open coast, in a cove at Burunof Point, we were frequently able to see these beasties "at home" before they had been disturbed, due to a combination of bright weather and clear unruffled water. *G. vertens* swim strongly and beautifully, but apparently never until forcibly detached. Normally they lie close to the seaweed, swaying with the va et vient ["comings and goings"] of the ground swell, holding on with two or three tentacles (the adhesive cells of which can attach even to the porcelain of preparing trays), with the rest of the tentacles stretched widely. When detached from their support by mechanical agitation (or by inordinate stirring up of the bottom silt, as in Puget Sound by motorboat propellers at ebb, or once when a gusty wind sprung up during an extreme low and left hundreds of the animals swimming about aimlessly) they swim in the proverbial *Gonionemus* fashion, coming rapidly to the surface upright, reversing, and drifting back to the bottom upside down, the tentacles widespread. It appears to me that the local *Gonionemus* may be in the many- generations-process of evolving from an active and free swimming form, to a *Haliclystus*-like animal, more or less permanently attached to the eel grass or algae. This would be an advantage in that it would enable these animals to colonize more open stretches of the violently agitated coastline, whereas they must be restricted now to well or completely sheltered regions. At present they seem to be limited to eel grass and certain algae, on suitable bottom where currents run fairly strong either from tides or ground swells, but well sheltered from the direct force of waves, and with clear and fairly oceanic water. However, since *Gonionemus* occurred, formerly at least, in such abundance at Friday Harbor that a single collector, entirely untrained in zoology, is known to have gathered more than 10,000 in a few days, and since we have never seen them in this fashion, it may be that there are environmental factors other than those we have evaluated.

Polyorchis penicillata A Agassiz. Occasional throughout the region. We averaged seeing possibly two or three per week, and there may be concentrated areas where they occur in abundance, but not, I suspect, in the great hordes of hundreds of individuals which occasionally characterize Monterey Harbor. The literature indicates that they were similarly indigenous to San Francisco Bay in the days preceding the extreme industrial pollution. The large Monterey specimens are in addition several times the size of the largest seen in the north.

Haliclystus stejnegeri Kishinouye. The sessile scyphozoan, may be found sparingly, occasionally in great beds, wherever there is eel grass, but occurs more frequently on *Enteromorpha*, in any case matching in color the surrounding medium, green, brown or purple. We found it very abundantly west of Port Townsend, Washington, July, 1930, sparingly at several places along the B.C. coastline, and at Sitka in August 1932, once at least on a leafy alga in a deep and sheltered crevice tidepool on the surf swept outer coast. Most surprisingly, small specimens turned up two springs in succession at Monterey, just back of our laboratory, on a stringy green *Enteromorpha*. As industrial pollution increased with the opening of the cannery season, they disappeared, but there must be an endemic area somewhere in this southern region, to keep them going. In collecting living specimens, I was surprised to see that they had certain independent power, not only of slowly gliding along the algae, but actually of swimming. The pads of adhesive discs are also exceedingly sticky; they can adhere to surfaces as smooth as glass.

The Family Bunodidae or Bunodactidae, tuberculated anemones, with its *Bunodactis, Urticina* and *Epiactis*, well developed on the Pacific, includes the largest and showiest anemones on the coast. Torrey (U.C. Publ. Zool., 111, No. 3, 1906) merged into *Bunodactis xanthogrammica* (Brandt) ranging from Unalaska to Panama, several animals widely differing in appearance and ecology, though probably structurally no more than wide variations in a series. In the confused history of this group, three species especially stand out, *xanthogrammica* and *elegantissima* of Brandt 1835, and *artemisia* Pickering in Dana 1846, all as *Actinia*. At least three formae are recognizable on the Pacific coast today, no doubt intergrading, but typically quite distinct especially north of Monterey.

Available descriptions of *artemisia* (as Fig. 15, McMurrich, Annals N.Y. Acad. Sci., XIV, No. 1, 1901) definitely fit the elongate, pale, and chiefly submerged form found at Comox, Fishermans' Cove, and elsewhere along the B.C. coast, at Sitka, and even, in one suitably sheltered region with *Paphia* and *Arenicola* near the Fishermans' Wharf, at Monterey, Calif. This occurs in the coarse sand or fine gravel, always in the low littoral, the pedunculate base attached to shells of living *Cardium* or to loose smooth stones well buried, the disc raised above the surface of the sand at flood tide, but retracted at ebb. The tentacles are long and thin, the disc brilliantly colored in greens and lavenders barred and marked with white, but the peduncle and base are dull green, gray, or even white. When first taken, completely retracted, the animal reminds one more of an elongate *Metridium* than anything else, especially since the tuberoulae are concentrated near the disc, and may not be noticeable in complete contraction. From 3 to 6 or 8" in length, not highly variable in color, form or habitat, this should be rated the rarest of the 3 forms, though still relatively abundant, and certainly obvious and easy to find. This may easily be regarded as, and probably is, a race of the following form, adapted to this specialized semi-burrowing habitat.

Its apparently nearest relative, a gregarious small form of the middle and upper littoral, may be seen to good advantage at Monterey, where it carpets the granite rocks under the canneries and under Fishermans' Wharf. Characteristically, this type averages 1/2 to 2" expanded disc, with tentacles intermediate in length between the long ones of *artemisia* and the stubby ones of the *xanthogrammica* form. The variable peduncle is typically no longer than its breadth, and from base to disc it may be covered with pebbles or bits of shell. The dull green may be relieved on the disc with bits of lavender, red or white. This is exceedingly a hard-boiled type; it tolerates cannery pollution, even prolonged stagnation (a rare trait in an open ocean form), and is highly resistant to narcotics. I have never seen this typically in inland waterways, although it rarely grows on violently surf swept rocks, but a similar, possibly identical anemone is found on the west shore of the Day Island, Tacoma. It occurs on the inner side of the fringing islands at Sitka. I have been thinking of this as the "*elegantissima*" form, but it fits the descriptions and illustrations (as McMurrich 1901, op cit.) only in a general way.

The form we have been calling *xanthogrammica* is quite definite. Solitary, in the sense that it never forms the coherent beds of *elegantissima*, occurring well down in the littoral and sublittoral, this large and handsome form, exceedingly difficult to preserve expanded, even through careful use of narcotics, is limited to the violently surf beaten rocks and clear water of the outer coast. The specimens on the exposed shores of the fringing islands at Sitka are almost identical with the sometimes larger individuals at Monterey. The finest I have ever seen, certainly 12" or more in diameter, occur in deep gullies and rock pools as at Pt. Lobos, Carmel Highlands and (especially) San Remo and below, on the central California coast. They are vividly blue-green in disc color, with transparent, short and stubby tentacles. South of the bold stretch of coast between Port San Luis and Pismo Beach, where this brilliant form is abundant, the representatives have a smaller and dingier aspect, until at Bird Rock (near La Jolla), and at Boca de la Playa (Lower California) they can only with difficulty be distinguished in size, color and habitat from large individuals of the *elegantissima* form. On the bold cliffs and in the cave region near Halfway House (Lower California) the typical form again appears, and

I suspect that physiography is the influencing factor. Dr. Moberg at Scripps Institute noted verbally that one of the large *xanthogrammicas* in the aquarium there had been much smaller and differently colored when first brought in.

Another bunodid anemone, the circumboreal *Urticina crassicornis* (O.F. Miller), ranges as far south as Carmel, but the northern specimens are more abundant and range relatively higher in the littoral. (The related *U. columbiana*, heavily beaded, occurs offshore in California.) This large and spectacularly red form was particularly noted at Lowe Inlet and at Fishermens' Cove, B.C., and on the outside of the islands fringing Sitka. The former expanded readily despite temperature variations, provided only the water was sufficiently saline; the latter remained contracted when put into water from Jamestown Bay (only a few miles inland, but possibly five degrees warmer), no matter how frequently changed or aerated.

In reading over this account, Dr. Fisher remarked the presence of two *Urticina* on the Pacific, and I was unable to recall which had occurred there in the north, having failed to differentiate the characters in the field. In the very distinct Monterey Bay forms, the littoral type has a smooth column streaked with green on a deep red background. The dredged specimens are heavily beaded with white on a vivid red column, and have been identified with *Urticina columbiana*. During the 1934 Washington trip I kept these distinctions in mind, and was able to differentiate two types occurring in the littoral at Clallam Bay (inland from Cape Flattery, on the Straits). In the first, the vividly beaded red column is streaked with dark sage green in irregular vertical stripes, or mottled with green or dirty white; the verrucae are not obvious, best seen in certain lights. In the second, the column was brilliantly blood red and uniform, lacking green or white blotches; very obviously beaded in regular encircling rows; the beads sometimes white tipped, but no nearly so spectacularly as in the Monterey deep water form. A specimen was seen earlier in the year with few, light and transparent beads. This second type proved hard enough to anaesthetize, but not nearly so difficult as the first, no single specimen of which was preserved expanded. Whether or not these two forms represent extremes of an intergrading series would have to be determined by quantitative examinations.

Pisaster-Evasterias. Previously I had been unable to understand the distribution of *Pisaster ochraceus* (Brandt) and *Evasterias troschelii (Stimpson),* the commonest littoral asteroids in the region covered. In their "effective" mutual ranges[5] they may sometimes be found side by side, but more often in regions predominantly populated by one or the other.

From an examination of probably inadequate data, the answer would seem to lie in the freshness-stagnation factor (but only in the northern region, see remarks in footnote with reference to the situation in Southern California). In very quiet coves, lagoons or inlets with only one entrance, *Pisaster* is not

[5] With the so-far-as-known sole exception of Newport Bay, Southern California (where MacGinitie verbally reports the presence of literally tons under the Newport Bay wharf, etc), *Evasterias* doesn't extend effectively south of Puget Sound in the littoral. Dr. Fisher records the finding of two only specimens in years of collecting. The writer's personal collecting of many thousand large starfish, chiefly near Monterey, has turned out not a single *Evasterias* unless an occasional specimen has slipped thru unnoticed. This is quite in contrast to its tremendous abundance in the quiet channels of the North. *Pisaster ochraceus* on the other hand ranges effectively at least from Southeastern Alaska on the north, to the Point Sal-Point Conception region; below there, as far as we have collected in Lower California, its occurrence as the sub-species *segnis* seems to be a matter of suitable topography.

frequently found; in rapids and narrows it is usually the only starfish. 75% or more of the large five rayed starfish on the perpendicular rocks predominant in eroded areas even inland, are *Pisaster*, the percentage on outside rocks is up to 99—strays always occur. Where the littoral starfish population of an inlet consists mostly of *Evasterias*, there is a likelihood that water circulation is slight. *Evasterias*, however, very frequently occur in chiefly subtidal belts, below the *Pisaster* zone, where the currents are definitely strong. This in spite of the fact that one of the formae of *Pisaster ochraceus*, *f. confertus* Fisher (P. 172, Bull. 76, Pt. III, 1930) is restricted to quiet inland waters, while the other two are chiefly open coast inhabitants. Even the most quiet water races of *Pisaster* obviously haven't acclimated themselves to conditions as still or even stagnant as have some of the *Evasterias*. However, it should be remarked that *Evasterias* is known to range the open coast of south central Alaska from the Pribilofs to Kodiak, although, nothing being said about depth in the listings, it may be sublittoral.

Cucumaria miniata (*Brandt*) has been mentioned already as occurring in huge belts in the low intertidal (3' below to 1' above mean lower low water) along the inland waterways, and on the semi-protected rocks at Sitka. It might be added that a typical colony of six or eight individuals (obviously strays) has been living for several years in a north facing cleft of rock near Carmel, on a fairly open coast line, but protected by fringing rock reefs. It would be interesting to know how far the pelagic larvae of these animals had drifted before they found attachment this far south of any other known colony. I presume. *C. miniata* may occur in Coos Bay, Oregon, although I have never seen any south of Puget Sound. There is always the possibility that the animals may continue south from there as a rare subtidal species.

If one were to attempt the difficult task of summing up the salient features of these diversified faunas, he might begin by remarking that the Pacific Coast, noted for its development of such groups as the nemerteans, bryozoa, asteroids, limpets, keyhole limpets, abalones, nudibranchs, hermit crabs and shrimps, lives up to its reputation in the north. The inside waters, especially, are famous for their rich pelagic associations. But a delineation of even the high lights of the region covered ought to start with the primary differentiation into inner and outer waterway inhabitants. The differences are so great they cannot be treated as one unit.

More detailed characterizations and faunal lists, too long for inclusion here, confirm the oft-repeated observation that the littoral fauna of the Pacific Northwest resembles more nearly that of Norway and Great Britain than it does that of the American Atlantic. Currents and physiography possibly account in part for this similarity. The natural topographic conditions are similar; British Columbia and Southeastern Alaska have fjords equivalent to those of Scotland and Norway. Both are eastern and precipitate shores of great oceans, and are characterized by rocky and sudden shores with great depths close in, and by tides of considerable amplitude.

The Sitka episode, particularly, crystallized a mass of observational data on the distribution of animals as related to variations in the wave shock factor. We had collected for weeks steadily in an environment free from ground swell and surf. Then suddenly, within a few miles, both appeared, we were again on the open coast. And more than coincidentally, the whole nature of the animal communities changed radically, more than it had in a thousand miles of inner waterways. The species were different, their proportions were different, they even occurred differently. The fauna of the surf swept rocks outside Sitka resembles that of the similarly exposed California coast nearly 2000 miles distant, more than it does that of similar type of bottom protected from surf, only three miles away. Obviously the pelagic larvae of both divisions must often colonize the so-closely-adjacent area of the other; some powerful environmental factor must sort out the sheep from the goats.

The faunal differences between the exposed and lee shores of Kyack Islands, and the rocky cliffs adjoining the mud flats at Jamestown Bay, all within a half hour's run by motorboat, are so considerable as to provide excellent textbook illustrations of three formations, tentatively describable as follows:

A. Communities subject to *direct wave shock* and/or surf (or a set of conditions tied up with these so that wave shock can be used as an index), on a bluff, unprotected and usually precipitous coast, with all gradations from this extreme to

B. Communities subject to indirect ground swell or surf, with wave shock modified by protection of headlands, outlying islands, reefs, gradually sloping beach, or offshore kelp beds. Communities of this type are very common on the lee shore of islands subject to heavy surf on the windward; often in protected tidepools on the exposed side. Ranging thus from the first category to

C. Fully protected communities. Animals *entirely* sheltered from surf and/or ground swell. Within this division, associations of windward and lee, or of current and stagnant shores could probably be realized. Strong tidal currents, however, may simulate wave shock conditions so closely as to limit their inhabitants to forms found elsewhere only on the open shore.

It would be easier to evaluate these communities if the gradations from one to another of the above arbitrary divisions were more abrupt. If the curve graphically representing the large means of these changes from exposure to shelter were in the form of steep steps followed by plateaus, instead of in a steadily ascending and probably straight line, natural divisions could be formulated. As it is, I suspect that the gradations, if at all discernible, are minute; and until someone comes along with a recording meter capable of measuring the footpound reading of surf, by the day, season and year, at various selected points along the coast, and possible logarithmic variations of this factor must remain unevaluated.

There are, of course, many obvious physiological distinctions between typical "A" animals and typical "C" animals, and a good many anatomical generalizations could be drawn to the same classification, earmarks, or indices, if one cares for them. The narcotic differential encountered in anaesthetizing inlet and open coast forms has already been mentioned. The structural adaptations utilized by open coast animals as protection from crushing or being swept from their attachment by wave shock are patent. No typical "A" animal can be without some morphological adaptation, heavy shell, resilient integument, powerful attachment organs, etc.

Theoretically, the larvae of "A" animals are always attempting colonization of "B" and "C" grounds, and vice versa, the tendency toward excessive production being universal and overcrowding being what it is, but there isn't sufficient food, or oxygen, or attachment space, etc., for all, and the animals adapted to their environment are more efficient in getting more than their proportional share.

An examination of these data suggests the validity of the following statement (I have to thank Mr. Joseph Campbell of New York for this clear and concise statement of the situation all of us observed and discussed):

> *Given, originally, an equal distribution of animals, the tendency would be (a) for inlet conditions to favor the survival of species better fitted to withstand a considerable annual and diurnal temperature range, occasional fresh water, stagnation, and the deposition of silt, and (b) for surf conditions to favor the survival of species better fitted to withstand persistent wave shock.*
>
> *The progressive assumption of supremacy in each habitat by the animals better fitted to withstand its peculiar limiting factors would progressively augment the survival difficulties which would have to be confronted by the others—for there would be added to the original geographic (i.e., physiographic) obstacles, an increasingly vigorous social obstacle. The better fitted would be giving battle, furthermore, in a terrain peculiarly favorable to themselves and peculiarly unfavorable to their enemies.*
>
> *Thus progressively, the tendency would be for environmental limiting factors to select out a society of species physically fitted to survive the*

imposed conditions, and for the consequently established society itself to reinforce its own supremacy by crowding out and eating exotic creatures adapted to other conditions.

Stated less concisely, the situation seems to be that the chief limiting factor in open shore communities is wave shock. The chief limiting factors of quiet water communities are probably temperature fluctuations resulting in the waters becoming unduly warmer than the ocean in summer and colder in winter; salinity fluctuations due to rain, to the influx of fresh water, or to evaporation by sun and wind, especially on flats at low tide; fluctuations in O_2 content sometimes involving stagnation; the occasional presence and deposition of silt; and probably other minor factors, important in their sum. Only very rarely is the lack of wave shock a limiting factor in inlets, but with some forms (*Tivela*, the surf-living Pismo clam) this may become a lethal factor.

In examining open shore communities, one notes the great development and variety of clinging organs, tough protective integuments and pressure-distributing shapes. So also the typically quiet water animals exhibit wide margins of safety in their physiological tolerances and lethal points regarding the four above-mentioned factors. Open shore communities are "adapted" to oceanic conditions, quiet water communities to continental conditions, highly modified in each case by shore influences.

Assuming a generalized distributing point for all generalized shore animals with pelagic larvae, dispersal will cast the larvae indiscriminately onto quiet and wave-agitated shores (currents and type of substratum being equal).

Physiological and morphological variation, normally present in marine animals with reference to the above-mentioned traits, will mean that some individuals are thrust into unsuitable environments in which they are at a disadvantage, and in which they may be unable to reproduce even if they succeed otherwise. Others will drift into environments eminently suitable, and there will be all gradations with as many balances struck. Natural selection, through killing off adversely varying individuals and mutant races not suited to their environments in the above factors, and in furthering individuals and races increasingly suited, favors, in the struggle for existence, tools that very accurately fit the environment, and blesses by a comparative degree of ease, the bearers of the successful tools. Animals physiologically suited to given conditions reinforce one another, so that compact colonies are formed of many coordinating species that can use for food any pelagic larvae swept in from other environments before such newcomers have a chance to settle and compete.

Looked at in this light, absence of wave shock may be the "permitting" factor which accounts for the presence intertidally in very quiet localities, of a good many species and probably some whole associations which occur along the California coast in very deep water only: *Terebratalia*, *Stylarioides papillata*—a bristle worm, *Ophiopholis aculeata* (exceptionally occurring littorally on the ocean coast but in limited numbers), the mason worm *Pectinaria*, etc. In scattering from the shallow sea bottom, colonists would normally radiate in all directions including up and down. Continuing to move up from a given dispersal point, and encountering no wave shock in the inland waters, they would migrate clear into the intertidal, being slowed down there and finally stopped somewhere along the gradient of increasing exposure to air and terrestrial conditions. On the open coast, at a point still well below the littoral, wave shock would slow down the progress of those not equipped to withstand it, so that only strays would be found, with the likelihood that the race would be unable to colonize there, unless specially adapted to survive this hazard.

Increasingly it seems to me evident that wave shock, working often indirectly through its allies physiography and type of bottom, is the most significant distribution factor for the Pacific coast littoral, lacking here as we do any effective temperature differential from Point Conception clear into Alaska. Wave shock, currents, type of bottom and shore physiography are tied up together in mutual cause-effect circles, so that while any one of these factors can be defined exclusive of the others, the causes and effects of one cannot be considered, sometimes cannot even be differentiated, except in their mutual reactions and relations. Stones or boulders, unless they be very large, cannot occur in the face of heavy surf; the only shore that bears up under continual wave impact is cliffy wall, solid roof or solid terrace. The type, slope and exposure of the shore are functions remotely of geology, but immediately of physiography, this in turn is regulated to some extent by the erosion of currents and wave action, which in their turn can operate to best advantage on jutting headlands and on friable material, a function again of physiography. A good many forces act from outside. The steady wind piles up the ground swell offshore, temperature and barometer variations in other parts of the world induce currents, the lay of the land is in part due to remote earth movements. But within a given region all act and react together and mutually. Together they produce a result which conditions the environment of the animal communities. And in this complex are tied up the primary ecological classifications, not all coordinate, the in- and epi-fauna, the opi- and endo-bioses of the Scandinavian investigators, Allee's associations of eroding and depositing shores, and the ideas of wave-swept, semi-protected and quiet water communities that have been presented here.

THE LEGACY OF A NATURALIST:
How Ricketts's Wave Shock Idea Helped Shape a Century of Shoreline Research

C. Melissa Miner, David P. Lohse,
Peter T. Raimondi, and John S. Pearse

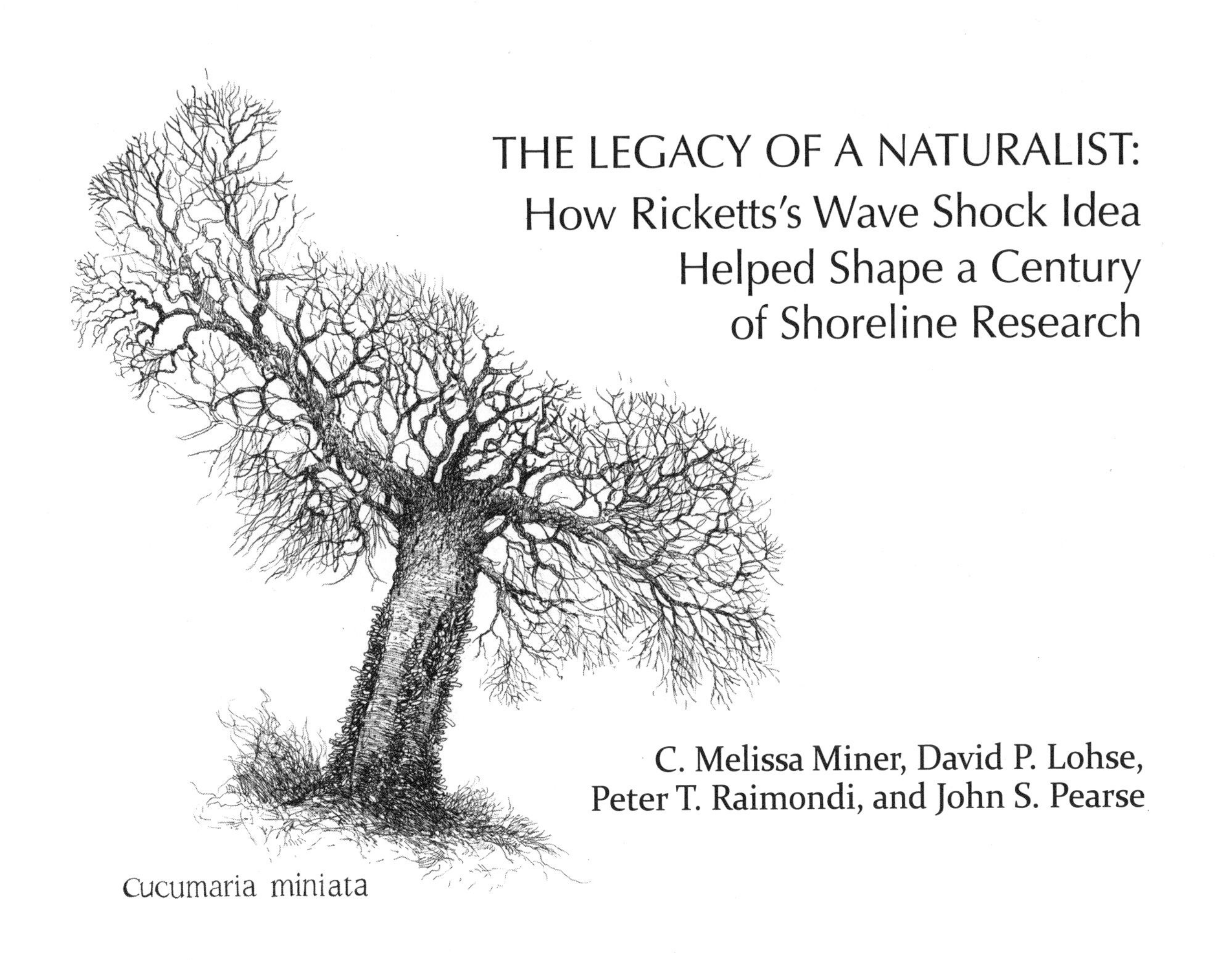

Cucumaria miniata

Specimens in the lab at Pacific Biological Laboratories. *(Courtesy of Nancy Ricketts.)*

NOTE: This essay includes excerpts previously published by Lohse DP, Gaddam RN, Raimondi PT (2008) in *Developing Wave Energy in Coastal California: Potential Socio-Economic and Environmental Effects*. Chapter 4: "Predicted effects of Wave Energy Conversion on communities in the nearshore environment."

OVERVIEW

When Ed Ricketts recounted his 1932 collecting trip to Southeast Alaska, he made special reference to the idea that "wave shock" could explain the strikingly consistent biological patterns he had observed along shorelines from California to Alaska. This idea—that wave intensity was a critical factor in determining both the types of organisms present in a community, and how these organisms were arranged within a stretch of shoreline—became the basis for many of the seminal ideas that emerged from coastal ecology in the decades to follow. Here, we summarize the key components of this family of ideas that relate to wave intensity—ideas ranging from the physical and biological challenges associated with wave energy to the changes that humans have caused in some areas by modifying the nearshore wave regime. Ricketts's detailed descriptions of the species he and Jack Calvin found in their surveys are also valuable as a historical record of what was present at that time, which can be used for assessing long-term community change.

INTRODUCTION

Ed Ricketts, a pioneer in the careful study of intertidal and nearshore communities along the western coast of North America, conducted his work in a manner that many present-day biologists look back upon wistfully. Done in an era when there was little concern about sampling methods, and conclusions drawn from observations did not require statistical support, Ricketts's studies instead relied on the power of observation, which to this day is often the most compelling measure for assessing change or identifying differences among areas. Indeed, many ecological studies can trace their beginnings back to some striking observation that merited further investigation.

Ricketts let curiosity and interest drive the focus of his surveys, which frequently resulted in exhaustive lists of intertidal animals (e.g., worms, anemones, cucumbers, sea stars, tunicates, sponges), but algae (seaweeds) were largely ignored. This can make current comparisons to his work challenging—often we are interested in whole community change, including algae. Moreover, there is always the question of whether species absent from historical records were actually present but just not recorded, and if those mentioned extensively were truly common, or simply the focus of intense searches. On the other hand, quantified studies performed today within defined areas (e.g., quadrats and transects), and with the replication needed to provide the statistical power to demonstrate change, often, by design, avoid nooks and crannies, under-rock habitats, deep tidepools, and organisms hidden within complex habitats such as mussel beds, or buried in sand and gravel. Consequently, they may exclude species that might have been seen with careful qualitative or semi-quantitative studies by observers such as Ricketts. Arguably the best approach is one in which natural history motivates a hypothesis that is then tested using a repeatable approach and produces data that can be used for comparison.

Perhaps one of Ricketts's most enduring contributions was based on observations made when he was aboard the *Grampus*, motoring first through the protected shores of Puget Sound, then north via the inner coastal waters of British Columbia to Southeast Alaska, where wave exposure varied substantially by location—all the while keeping the semi-exposed shores of Monterey Bay in mind. During this trip, he concluded that the striking variations in species composition and abundances in intertidal communities present along this stretch of continent were primarily due to differences in wave energy. This conclusion was consistent with the literature of his day, with which he was very familiar (e.g., Verrill and Smith 1874, Flattely and Walton 1922, Johnson and Snook 1927—all marked as "indispensable" in Ricketts and Calvin 1939). He was particularly impressed by what he called the "Sitka episode" in the "Wave Shock Essay":

> *We had collected for weeks steadily in an environment free from ground swell and surf. Then suddenly, within a few miles, both appeared; we were again on the open coast. And more than coincidentally, the whole nature of the animal communities changed radically, more than it had in a thousand miles of inner waterways.*

That led to his proposal of three types of rocky intertidal communities, those subject to (1) direct wave shock, (2) wave shock dampened by protection from headlands, outlying islands, reefs, gradually sloping beach, or offshore kelp beds, and (3) no wave shock. Ricketts and Calvin's famous and influential book *Between Pacific Tides*, first published in 1939, was completely organized around these three community types: "Protected Outer Coast," "Open Coast," and "Bay and Estuary," with the addition of a section on "Wharf Pilings."

Investigating how wave exposure shapes intertidal communities continues to be one of the most commonly explored areas in marine ecology. Wave exposure has been identified as the primary driver of observed patterns of difference between exposed versus protected shores, and modern researchers have broken this very general observation into smaller pieces, investigating more manageable, testable components of this broad idea. Below, we attempt to summarize some of the key ideas that have been explored at both the physical and biological level in the quest toward gaining a better understanding of Ricketts's general "wave shock" hypothesis.

DEFINING "WAVE EXPOSURE"

In his descriptions of nearshore habitats, Ricketts used the categorical terms "quiet" and "agitated" to group areas, based on the amount of wave energy they experienced (wave climate). While "quiet" has morphed into "protected," and "agitated" into "exposed," subjective terms are still widely used today to classify coastal areas and to explain observed differences in community composition (e.g., Blanchette et al. 2007). However, as Ricketts (2020) points out, anyone attempting to pigeonhole a collection of sites into two or three "exposure" categories quickly realizes that wave energy is a continuous variable, and sites should instead be arranged along a continuum of wave energy (Lindegarth and Gamfeldt 2005). This idea is particularly important because, as Ricketts (2020) suspected, the composition of a given biological community can change dramatically given a slight change in wave climate (Lindegarth and Gamfeldt 2005; Burrows et al. 2008). In other words, community response to wave energy is non-linear.

A good example of a non-linear relationship between wave energy and community structure comes from the kelp forest communities along the coast of central California. Kelps are important species not only because they are the main primary producers in many nearshore communities, but because they provide structure to the water column that is fundamentally important to the diversity of kelp communities (Graham et al. 2008). Along the California coastline, the giant kelp, *Macrocystis pyrifera*, tends to be found in lower wave energy environments, while bull kelp, *Nereocystis luetkeana*, occupies more exposed habitats (Foster and Schiel 1986; Graham et al. 2008; Schiel and Foster 2015). Observations suggest that the switch from one kelp species to the other appears to fit the non-linear threshold scenario. That is, a relatively small difference in wave energy can have a pronounced effect on which species of kelp is present and, therefore, the composition of the resulting community (Hagerman and Bedard 2004).

A NOTE ABOUT SCALE

When researchers quantitatively investigate how wave climate varies along coastal habitats, an important consideration is the scale at which measurements are taken. Obvious patterns of difference exist on the scale of kilometers between exposed versus protected shores, but even at small scales—tens of meters or even centimeters—we see differences in wave run-up (the distance a wave travels up the shore beyond the still water level) and splash due to differences in topography and orientation to waves (Denny et al. 2004). The same considerations must be made for temporal scale. We know that wave height and direction vary somewhat predictably by season, and that major storm events (which bring large waves responsible for significant levels of disturbance) typically occur in winter. However, a substantial amount of variation surrounding these events must be factored into any generalizations about how wave climate influences coastal and nearshore communities (Denny et al. 2004). Because substantial variability exists in wave climate at both the spatial and temporal scale, it is important to define the scale of interest when drawing conclusions from observed patterns.

PHYSICAL PROCESSES AFFECTED BY WAVE EXPOSURE

Organisms that live in wave-swept habitats must deal with a variety of physical challenges. The hydrodynamic (fluid in motion) force resulting from a wave crashing on the shore can be quite large, and waves often carry material suspended in the water column. Thus, an organism in a wave-swept habitat runs the risk of being ripped from the shore, abraded by sand particles, or crushed by logs (or other water-borne objects). However, the same waves that might cause destruction can also prevent desiccation and can moderate nearshore temperature. Moreover, wave-beaten shores receive more energy from breaking waves than from the sun, and much of this energy enhances productivity by bringing in nutrients, removing debris and waste products, and inhibiting the establishment of competitors and predators, analogous to enhancing terrestrial productivity by human farming practices (Leigh et al. 1987). Just how well an organism is adapted to meet these challenges and opportunities can influence where along the shore it lives and its chances of survival.

EMERSION TIME

The tidal cycle dictates that intertidal organisms spend part of their time exposed to the air (emersion) and part immersed in the water. Typically, locations higher on the shore spend more time in air and less time submerged than those lower on the shore. Because species differ in their ability to tolerate this gradient, they are not distributed uniformly throughout the intertidal zone. Instead they are found in bands along the shore, a phenomenon known as vertical zonation.

Although intertidal zonation patterns are largely determined by the tidal cycle, wave exposure also plays an important role. Specifically, because of wave run-up and wave splash, incoming waves extend the upper boundary of the intertidal zone above that which is set by the tidal cycle. In general, the larger the wave, the greater the run-up, and the higher the intertidal zone extends on the shore. Thus, as Ricketts and Calvin (1939) mention, the zone occupied by each species tends to be broader and more diffuse, and the upper levels are located higher on the open coast, which is exposed to large waves, than on protected outer shores or in bays and estuaries (Figure 1, see next page).

TEMPERATURE

In general, sea surface temperature (SST) decreases with increasing latitude. Thus, how well a species can tolerate warm/cold temperatures will strongly influence its biogeographic distribution along the coast. An important factor that influences local SST is the depth of the mixed layer. Since the heat generated by the sun is shared throughout the mixed layer, the shallower the mixed

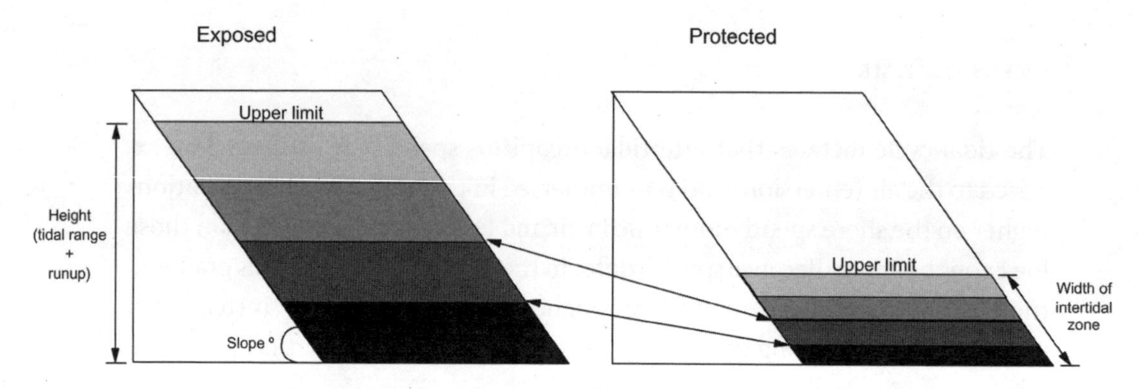

Figure 1. Comparison of zonation patterns at wave-exposed and wave-protected intertidal sites. Note that each zone is narrower and located lower on the shore at the protected site. (From Lohse et. al 2008)

layer, the less volume of water is being heated, the warmer the SST. The depth of the mixed layer is generally shallower in lower wave energy environments, which in turn affects species composition (Largier et al. 2008). Protected areas, especially bays and estuaries, which are generally shallow and have little mixing, can experience much more extreme temperature fluctuations than open coast, wave-influenced habitats. Consequently, species in protected habitats can be expected to tolerate a wider temperature range than those on the open coast (Ricketts 2020).

DISTURBANCE

A disturbance is an event that is often physical (rather than biological) and that indiscriminately affects individuals. In the nearshore environment, disturbances are usually the result of large waves striking the shore. For instance, large, storm-generated waves can cause boulders to flip or roll (e.g., Sousa 1979; McGuinness 1987a, b), can rip organisms from their substrate (e.g., Dayton 1973; Paine 1979, 1988; Paine and Levin 1981), or cause water-borne objects, such as logs and rocks, to strike the shore (Dayton 1971; Shanks and Wright 1986). This removal of individuals creates patches of open space, which can in turn, expose the remaining individuals to more hydrodynamic force (Denny 1987; Bell and Gosline 1997). Once a patch is formed, it is not uncommon for subsequent waves to enlarge it (e.g., Dayton 1971; Denny 1987; Guichard et al. 2003), thereby resulting in more mortality. In kelp forests this can happen when a disturbed kelp plant entangles itself around its still attached neighbors, causing them all to get ripped from the substrate (Rosenthal et al. 1974). Although many studies have examined the role of disturbance in communities, there is surprisingly little data on whether the frequency and size of disturbance events varies depending upon the amount of wave exposure. What is available suggests that the size of disturbance events increases with increasing wave exposure (Paine 1979; Menge et al. 2005). This is supported by evidence that more wave-related disturbances appear to occur during winter, when wave energy is high, than during summer (e.g., Paine and Levin 1981; Menge et al. 1993; Blanchette 1996).

SEDIMENT TRANSPORT AND DEPOSITION

The movement and deposition of sediments in the nearshore environment are strongly affected by wave energy. In general, the amount and size of the sediment suspended in the water column is positively related to wave energy. At some rocky shore sites, sand is an important agent of disturbance (e.g., Taylor and Littler 1982; Menge et al. 2005). Depending upon the velocity of the water, sand can either act like an abrasive that scours organisms with each passing wave, or settle out of the water column and bury organisms attached to a rocky reef (Littler et al. 1983; D'Antonio 1986; Menge et al. 1994). Because organisms differ in their ability to tolerate these processes, the distribution of species both within (e.g., D'Antonio 1986) and among sites (e.g., Schoch and Dethier 1996) can be influenced by sand. On the other hand, as pointed out by Ricketts (2020), fine-grained silt can be washed into shallow bays and estuaries and remain in suspension, blocking light and fouling benthic organisms, and selecting against those requiring water with little sediment as is more typical of open coast habitats.

BIOLOGICAL PROCESSES AFFECTED BY WAVE EXPOSURE

In addition to its effects on physical processes, wave exposure influences how organisms interact with one another, how they eat and grow, and their physical characteristics. Two of the most important interactions among coastal organisms—competition for resources (e.g., space, food/nutrients), and predation (animals eating other organisms)—can be strongly affected by wave exposure.

COMPETITION AND PREDATION

Studies have found that the ability of predators to search for and feed on prey is reduced in high wave energy environments (Menge 1978; Sebens 2002). Therefore, rates of mortality from predation tend to be higher in wave-protected areas than wave-exposed areas (e.g., Menge 1976; Boulding et al.

1999; Robles and Desharnais 2002; Robles et al. 2010). Similarly, differences between exposed and protected sites in both diversity and abundance of seaweeds have been linked to differences in grazing pressure and nutrient availability (Nielsen 2001, 2003). Because the intensity of competition depends upon population size, competition tends to be less important where rates of predation are high. Thus, reduced wave energy could decrease the relative importance of competition. For example, because predation on the mussel, *Mytilus californianus*, by the sea star, *Pisaster ochraceus*, is more intense where wave energy is low, *M. californianus* is less abundant and occupies less of the shore in wave-protected areas than in wave-exposed areas (e.g., Robles and Desharnais 2002; Robles et al. 2010). Since *M. californianus* is a better competitor for space on rocky shores than many other organisms (Paine 1966, 1974), and a foundation species supporting many other species sheltered within mussel beds, any changes in its abundance can have important consequences for the structure of the entire community.

PHENOTYPE

Many organisms modify their size or shape (phenotype) to be better equipped for threats such as dislodgement, desiccation, and predation—threats that all vary in intensity along a wave energy gradient. Numerous species have distinct phenotypes in areas of low wave energy versus high wave energy that are a direct result of the challenges each habitat presents. Often, obvious size differences between individuals of the same species can be directly related to the threat of dislodgement by waves in exposed versus protected habitats. Small, streamlined organisms have a lower risk of being dislodged by wave force than large organisms with complex external structure. For example, Gaylord et al. (1994) found that the maximum sizes of three algal species were constrained by the acceleration and drag forces associated with wave energy.

Size and shape of the original "keystone predator," the ochre star, *Pisaster ochraceus*, is also affected by wave force. On protected shores, stars are heavier, with plumper arms than their open coast, wave-exposed counterparts, who have lighter bodies and narrower arms that likely minimize lift and drag resulting from breaking waves (Hayne and Palmer 2013). This response by *P. ochraceus* to wave energy has been shown to be phenotypically plastic—stars moved from a protected site to an exposed site will become lighter and develop thinner arms, and vice versa. Ricketts (2020) noted that one phenotype of *P. ochraceus* (formerly called *P. ochraceus forma confertus;* Fisher 1930), was restricted to calm inland waters, while other forms were found chiefly in open coast habitats.

PRODUCTIVITY AND GROWTH

The relationship between wave energy and productivity/growth is complex, and varies greatly across the wave energy spectrum. Quiet inland waters—which are warm and shallow, with abundant nutrients—can be enormously productive and are important nurseries for fishes and crustaceans (Beck et al. 2001). Moderate to high wave action on the outer coast can reduce predation and competition and facilitate nutrient uptake and photosynthesis in seaweeds such as the sea palm, *Postelsia palmaeformis* (Leigh et al. 1987), and deliver more food to filter feeders such as mussels (Blanchette et al. 2007), but at very high wave energy levels, these benefits are outweighed by increased disturbance. Reed et al. (2011) illustrate this idea with their study comparing net primary production of giant kelp beds in Southern versus Central California. Southern California kelp beds had consistently lower levels of nutrients and higher numbers of grazers, yet the average yearly net primary production was twice that of Central California kelp beds. This difference was attributed to significantly higher levels of wave disturbance in Central California, which resulted in greater removal of giant kelp.

GENERAL PATTERNS IN SPECIES COMPOSITION AND ABUNDANCE

Because species differ in their ability to live in fast-moving water, the composition of the community can differ even between adjacent sites, if the sites differ in exposure. For example, the area of the shore dominated by the sea cabbage, *Hedophyllum sessile*, in more wave-protected sites, is occupied by the flat pompom kelp, *Lessoniopsis littoralis*, in higher wave energy areas (Dayton 1975). Similarly, the tough but flexible sea palm, *Postelsia palmaeformis*, is common in high wave energy areas, but absent from more protected locations (Paine 1979; Nielsen et al. 2006). In subtidal kelp forests, as already mentioned, the giant kelp, *Macrocystis pyrifera*, is found in wave-protected areas, while the bull kelp, *Nereocystis luetkeana*, occurs in wave-exposed locations. Since species such as *M. pyrifera* and *H. sessile* are canopy species that provide structure or shelter to other species, their distributions affect the distribution of many other species in the community.

The trophic structure of communities has also been shown to vary with wave action. In one study of intertidal communities in South Africa, autotrophs (algae and plants), filter feeders (e.g., barnacles and mussels, which filter out food from the water column), and invertebrate predators (e.g., crabs and sea stars) were more abundant on wave-exposed shores, as compared to protected shores, and grazers (on algae) were more prevalent on sheltered and semi-exposed shores (Bustamante et al. 1996). In this same study, overall biomass was found to be higher on exposed shores than protected shores, but sheltered and semi-exposed shores were more diverse. Similarly, Zabin et al. (2012) found that species richness generally increased along a wave-exposure gradient on the northern shores of Monterey Bay, although this gradient was confounded by distance from an urban area.

A 2012 survey of species on the sheltered versus exposed sides of the Kayak Islands near Sitka, clearly an important site in the formation of Ricketts's wave-shock hypothesis, also found higher species richness on the exposed side of the island (86 species) as compared to the protected side (70 species). In the same survey, 39 species were found only on the exposed side of the island, and 23 were unique to the sheltered side (47 shared between areas). It must be noted that this was a rapid survey, and it is likely that the complete list of species would be longer given a more extensive search effort.

INFLUENCE OF HUMAN ACTIVITY ON WAVE ENERGY

For decades, humans have modified how wave energy affects shorelines. Enhancing natural harbors with breakwaters, dredging to provide safer entry points for ships, and armoring shorelines (e.g., seawalls, rip rap) are just a few examples. These modifications affect where and how waves break, and the transport and deposition of sediment, which in turn creates changes in neighboring nearshore marine communities. Now, growing interest in converting wave energy into electricity has led to the concern that wave energy conversion devices might significantly alter the wave climate of the high wave energy environments where they are being installed.

HISTORIC COMPARISONS OF SPECIES DISTRIBUTIONS AND ABUNDANCES

As society becomes increasingly aware of impacts to nearshore communities that can result from human activities, such as the types of shoreline modification described above or sea level rise due to climate change, we have realized the importance of having "baseline" data that allow for assessment of community change. Of particular value are historic studies documenting species distributions, which can be repeated and used to identify long-term shifts in the distribution and abundances of species.

In the summer of 1947, three University of California Berkeley graduate students designed a study that would be "a quantitative measure of the effects of exposure to various degrees of wave action and splash on the pattern of vertical distribution" (Hand et al. 1947). Thus, their study was a direct test of the ideas about the influence of wave shock on species distribution, mentioned in Ricketts & Calvin (1939) and more fully developed in Ricketts's "Wave Shock Essay," which is reproduced in this book. Perhaps because the essay is primarily directed at large-scale community composition, rather than the small-scale zonation that was the focus of the students' study, their report includes no direct reference to Ricketts's earlier observations. Yet, precisely because Ricketts's "Wave Shock Essay" is based mostly on southeast Alaska, their confirmation of the same patterns in his home territory of Monterey Bay is particularly satisfying. Incidentally, all three students went on to become prominent zoologists: Cadet Hand was a professor at UC Berkeley and the founding director of the Bodega Marine Laboratory of UC Davis; Donald Abbott was a highly esteemed Stanford professor in residence at Hopkins Marine Station; and John Davis was the Reserve Director of the Hastings Natural History Reservation (UC Natural Reserve System) and Research Zoologist, Museum of Vertebrate Zoology, UC Berkeley.

In the Hopkins study, Hand, Abbott, and Davis chose four sites to compare, two of them more exposed to the force of waves, High Rock and Hot Rock, two of them more sheltered from wave energy, Murphy's Rock and the enigmatic Snad Rock. The students counted six common high-zone taxa on vertical transects: two littorine snails, two limpets, and two barnacles. All six extended higher in the exposed sites (Fig. 2) than in the sheltered sites, a striking validation of Ricketts's wave-shock insight. Exactly sixty years later, these same four sites were relocated by John, Vicki, and Devon Pearse; the transects were surveyed in 2007 and re-surveyed biannually over several years. The results confirmed the wave-shock-related patterns observed by Hand et al., and more strikingly, they broadly reflected the rise in sea level over the intervening six decades. However—as so often happens—the data displayed new puzzling twists. Not all six animals changed in level equally. The total abundances of all of them were greatly diminished. And molecular genetic analysis has revealed that the species name used by Hand et al.

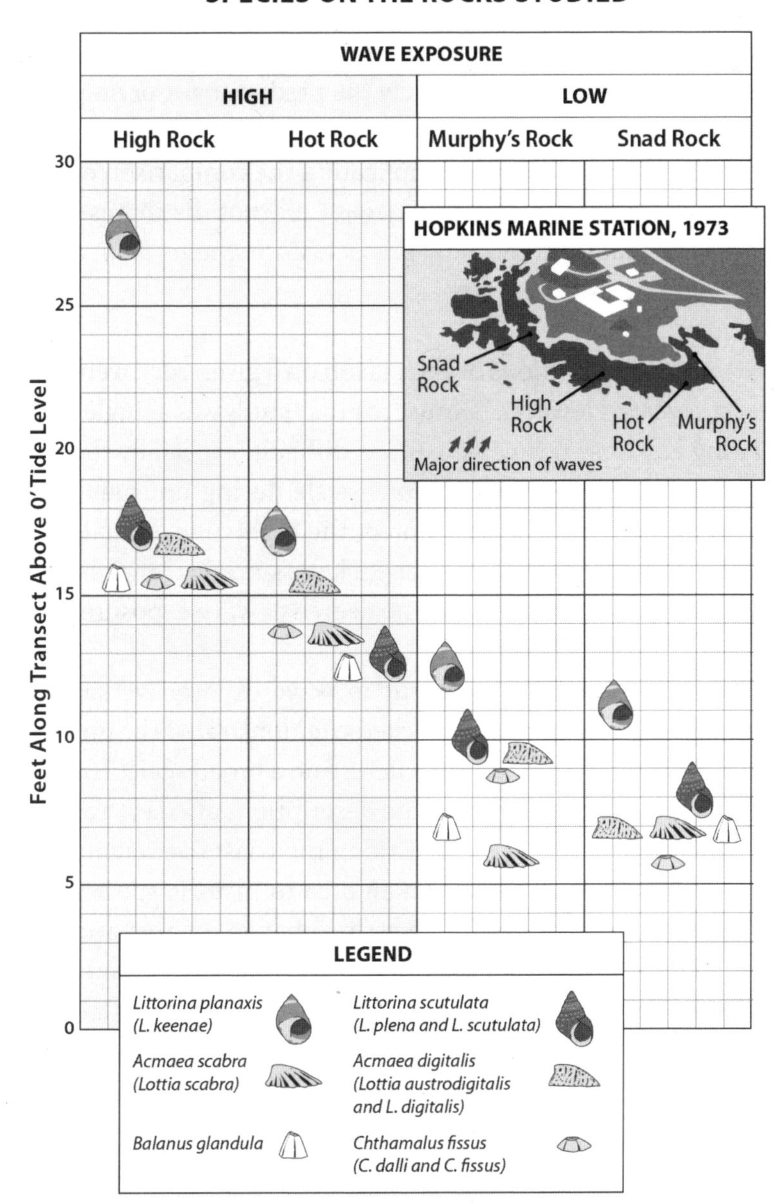

Figure 2. Position of the highest individual of each species on the four rocks studied. From Hand, Abbott, and Davis, 1947. Insert is a map of the shore at Hopkins Marine Station showing the four study sites based on an aerial photograph taken in 1973 (figure redrawn by Alison Kendall Swearingen).

for one of the limpets confounds two closely related species, *Lottia digitalis* and *L. austrodigitalis*. In 1947, *L. digitalis* was likely the predominant or only limpet present, but we know that their respective northern versus southern range distributions now overlap at Hopkins, complicating the comparison of data from the two eras. In addition, Hand et al. (1947) did not distinguish the two species of barnacles that were certainly present, *Chthamalus dalli*, a northern species, and *C. fissus*, a southern species.

In the 2007-2018 surveys, the distributions of each taxon on a given rock were generally consistent across years. However, because the 1947 study was a single-year snapshot, evaluating any long-term changes was difficult, especially the decrease in abundances. Did the early populations settle during unusually favorable conditions? Was there more fog to protect the high zone animals from desiccation? What is clear, however, is that Ricketts's original insight was correct: vertical zonation is strongly tied to the degree of wave exposure.

By documenting the types of organisms present in wave exposed versus protected areas during the 1932 collecting trip, Ricketts demonstrated the importance of wave shock in shaping community structure on a broad scale. His carefully recorded lists of organisms and associated descriptions also serve as important records of where species were located along the coast over eighty years ago. Comparing current species present in an area to historic species records is another approach scientists use for assessing change. In 2012 and 2018 a group composed of researchers from the University of California, Santa Cruz and the Sitka Sound Science Center resurveyed two areas visited by Calvin and Ricketts using a biodiversity sampling approach, producing species lists that could be compared to their historic lists (see table pages 168-172), while also providing estimates of species abundance and distributions, which could be used to better assess future community change. Comparing lists of species across time can be problematic, especially comparisons such as ours, where methods are vastly different. Calvin and Ricketts did thorough searches, moving rocks, looking deep in crevices, and in many cases hunting intently for species of particular interest. By contrast, our species lists were compiled from surveys done in a systematic, quantitative way. Species that are absent from our current list might actually be present but missed by our sampling approach. In addition, there is an inherent "miss rate," particularly for rare species, that must be considered when comparing survey results. For the 2012 and 2018 surveys, which were done using exactly the same approach, five species were found in 2012 but not in 2018 (excluding species unique to the Kayak Islands Outer Coast, which was not sampled in 2018), and one species was found only in 2018. It is likely that these rare species were not truly absent from surveyed areas, but rather were just not captured by our sampling.

Examples of species recorded by Ricketts but not found in our biodiversity surveys include several species of anemones and crabs that were of special interest to Calvin and Ricketts, and *Cucumaria miniata*, a sea cucumber often found in tidepools, which were avoided by our surveys. Another species not found in 2012 or 2018 is the northern abalone, *Haliotis kamtschatkana*. Abundance of the northern abalone declined severely with the re-introduction of sea otters into the Sitka area (J. Straley, pers. comm.), and while abalones still occur subtidally (T. White, pers. comm.), they are certainly less common than in 1932. Despite the limitations in making comparisons between the

1932 and more recent surveys, there was significant overlap in species documented, and all serve as permanent records of what was present at those points in time.

SUMMARY

It has been argued that on wave-swept shores, wave force is perhaps the most important physical stressor in intertidal communities (Denny 1995). In the quest to better understand how wave force affects both the physical and biological processes that shape community structure, researchers' questions have become increasingly focused and the approaches to answering them more mechanistic. However, the root of all these studies remains the same as that in Ricketts's time: questions arise from observations made while we are out exploring the shoreline. By continuing to investigate the processes responsible for the patterns observed by Ricketts nearly a century ago, we will increase our understanding of why these particular patterns exist, how they could be altered due to continued coastal modification and development, and the changes that will occur with anthropogenic global warming.

Dedicated to our friend and colleague, Dr. Rafael D. Sagarin, who shared Ed Ricketts's love of shoreline creatures, his enthusiasm for sharing his knowledge with others, and his sense of humor and creativity in thinking about how the world works.

RICKETTS, CALVIN, and THE CHARISMA OF PLACE

John Straley

Cannery Row, circa 1945. (*Photo by George Seideneck, Pat Hathaway Collection, CV# 1972-012-0053.*)

If I understand what Edward Ricketts, Jack Calvin, Sasha Calvin, and Joseph Campbell were contemplating as they worked their way up the coast on board the *Grampus* in 1932, simply put, it was this: *place shapes life.* If this is so, what clues do these words give us to the lives that must have been found in the bustling little community of Monterey, California, the place where it all started?

> *Cannery Row in Monterey in California is a poem, a stink, a grating noise, a quality of light, a tone, a habit, a nostalgia, a dream. Cannery Row is the gathered and scattered, tin and iron and rust and splintered wood, chipped pavement and weedy lots and junk heaps, sardine canneries of corrugated iron, honky tonks, restaurants and whore houses, and little crowded groceries, and laboratories and flophouses. Its inhabitants are, as the man once said, "whores, pimps, gamblers and sons of bitches," by which he meant Everybody. Had the man looked through another peephole he might have said, "Saints and angels and martyrs and holymen" and he would have meant the same thing.* – Cannery Row *by John Steinbeck*

Clearly, this was, through the eyes of John Steinbeck at least, a culture of sensual beings, who valued freedom over status. This was a culture newly built full of flaws, mistakes and promise. This was the American West.

I first read *Cannery Row* in the summer of 1967. I was staying in my parents' home in Atlantic Highlands, New Jersey. I had mononucleosis and I could not go to my summer job packing mules in the Cascade Mountains, but stayed in bed most of the summer reading, and I took the bus into New York City once every two weeks to have my blood drawn. It was a hot, drowsy summer. I also read *The Great Gatsby* and *The Adventures of Huckleberry Finn.* Even though I was to go to a prep school in the fall, and I was trading my cowboy boots and Western hat for an eastern coat and tie, I knew that after reading those books, particularly those opening lines of *Cannery Row*, that I was a Westerner. What's more, that I was in love with the West. The environment and culture of the American West created the boundaries of my natural range.

The west coast—from Baja California to Sitka, Alaska—was Ed Ricketts natural range as well. Ricketts was a man naturally suited to breaking the confines of conventions: he had wandered across the country and he was an adventurer. He had stood hip deep on the edge of the Pacific Ocean and he had meditated deeply on the taxonomic organization of animals and of his own place in the universe. Like Nick Caraway and Huck Finn, he had run west and he belonged there. He not only ran away from the east he ran towards the wildness.

"Literature of Place" of course, can be written about anywhere. It requires only a collaboration of a mind and an environment. Steinbeck's opening was particularly evocative and it propelled many readers towards an interest in the place and in the characters. "The boys" and Doc Ricketts, the girls of Flora's whorehouse and Chong's Grocery, have now blurred into larger than life, semi-mythological characters who no longer are bound by their historical record. They belong more to Steinbeck's romantic imagination than they do to themselves anymore.

But the place itself, Monterey, between the wars and just after the Second World War...there was something magical about it, not just the smell of the trees and the salt sea but the people who gathered there. Just as there was, and is, something magical in the city at the northern end of the range that Ricketts's partner, Jack Calvin, settled in—Sitka, Alaska, where the rawness of the west was, and still is, captivating to the adventurous soul.

Jack Calvin, who was the skipper of the *Grampus* and co-author of *Between Pacific Tides*, homesteaded in Sitka in the late 1920s. He was at home in the northern quality of light, this habit, this nostalgia. What's more, he saw in abundance in southeastern Alaska what was quickly disappearing in northern California: wild country.

Many years after John Steinbeck died, I wrote in *The Curious Eat Themselves*:

> *I love Sitka. There are eight thousand people, twelve miles of road, and two main streets. It had once been the capital of Russian America. To me it's a town full of mystery and wildness. It's so crowded by the wildernesses of steep mountains, thick woods, and ocean that a person can have the sensation, on the same afternoon, of either floating away or taking root.*
>
> *Great upwellings of ancient basalt and three-legged dogs are on the streets. There are gulls and murrelets. Cormorants lift their wings to dry their inky black feathers in the sunlight. Puffins with colored tufts like Gypsy scarves. Humpback whales feed on the herring that are feeding on the effluent from*

the pulp mill. Pickup trucks and Subarus. Everyone on their way to a meeting or softball practice.

Four kids with canvas jackets and earrings, with their hats on backwards, standing in the doorways near the Russian cathedral, looking bored. An old man walking outside the Pioneers Home in the middle of town, wearing pistols holstered on the outside of his pants. Occasionally a brown bear in the cemetery or a deer swimming in the harbor. The cathedral and the jagged, ancient mountain like a background for all our arguments. Priests, tourists, loggers, bureaucrats, fishermen, even an amateur whore or two, and one full-time private investigator.

I am not John Steinbeck. But I like to think I walked in his footsteps. I like to think that reading his words imprinted on me, during that drowsy summer of reading, the first intense love of place that I have kept alive all my life. So that when I came to Sitka back in 1977, even though I hated the rain and even though I hated the fact there were no horses anywhere to be seen, I could tell there was adventure and beauty to be had if one were game enough. Some places have charisma to shape stories and draw storytellers to them. Sitka is such a place.

Jack Calvin recognized this. At first the temperate rain forest and Pacific Grove's semi-arid coast line with its eucalyptus and shore pine forest would seem to have little in common. Yet both have similar abundant intertidal life of the north Pacific Basin; kelp forests and sea otter populations, abalone, (though different kinds) abound. Grey whales migrate past Pacific Grove and Sitka every year; humpbacks, orcas, fin whales, sperm whales, and blue whales are found offshore. Little silvery fish hold down the basis of the food chain. In Monterey, traditionally, it was sardines; off Sitka, it is herring.

Monterey can be described as a largely Mediterranean climate, and it drew Italian, Mexican and Spanish fishermen; wine and vegetables were grown inland from the coast and to the north. Remarkable food and cultural mixing occurred with the Chinese and Filipino men and women who were brought in to work in the canneries. In southeastern Alaska, we have a more northern climate with cod, halibut, and herring, which brought Scandinavian fishermen. The canneries recruited directly from Seattle and management often encouraged separate-language work crews, and separate dorms and dining facilities, to thwart mixing and labor organizing. So Chinese, Filipino, Native, white...while they did mix and intermarry, we didn't get the interesting widespread interweaving of cuisines you see in many border communities.

Remember, this was the early thirties, and Calvin was an intellectual, the co-author along with Ed Ricketts of the seminal *Between Pacific Tides*. To my mind, if Ricketts was the philosopher of the group, Calvin was the writer and the doer. He had paddled a canoe with his new wife from Seattle to Sitka and had written about it for *National Geographic*. When the art scene in Monterey grew too crowded and divisive, Jack needed more room. To my mind, Ricketts and Calvin were the forefathers of the nature philosophers to come: Gary Snyder with his intense scientific and philosophical interests, and now of course, all the scores of young people with backpacks and their degrees in environmental science, coming to Alaska to see the last of what's left.

And Alaska does have room for them all. Russell Banks, the American novelist and author of *Cloud Splitter*, has said that the two currents of the American narrative are race and space. When looking at Steinbeck and *Cannery Row* you can see this. "The Boys" are not white, and they are also not confined. They are if anything (and I have to admit in a modern context, condescendingly) children of nature. Their lives are free. They have room to wander and so too, Doc Ricketts. He travels along the edge of the great Pacific Ocean and wades out into it to collect its mysteries. He has room enough. He is not confined.

So too, Ed's daughter Nancy, who lives in Sitka to this day. I've asked her why and she tells me it's because Sitka reminds her of the Pacific Grove of her childhood. "The fish plants and the people working there, the music festival in the summertime. My father would have loved that. The gulls and eagles. Mostly though it's all the people here in their rubber boots who just want to get out on the water and out into it [the wild country]. That's what reminds me of the old Monterey. Not now of course."

Sitka sits on the outside coast, in the middle of the Tongass National Forest—the largest national forest in the United States. Waves roll from thousands of miles across the ocean to crash on the beaches. There's a small boat, longline and troll fishery for black cod and salmon based in Sitka. In a busy summer there are about nine thousand residents. Once a year there is a chamber music festival where musicians come from all over the world to play here for a month. They accept no payment other than their living expenses. One has built a home on an island here. Writers come for retreat and musicians come to play. Writers, artists, musicians of national reputation call Sitka *Home*.

They live here for the same reason that the Tlingit people settled here thousands of years ago and for the same reasons Jack Calvin did in the twenties. Standing on the beach fringe on a summer day in Sitka, one can smell the sunny tide flat and the clam squirt of a pristine beach, the fresh breeze of a cold ocean, blue-green water with waves cresting off the rocks, and only the smell of water, kelp, and spruce trees on the air. The ocean for Ricketts and the great waves, like the prairie for Aldo Leopold, made him consider the infinite in the subset of his tiny study area. The fresh wind held the possibility of mortality. In Alaska, death and food are always at hand on the beach. Calvin knew all about the power of northern waterways and shorelines. He built or helped build several houses on the beach. He brought his friends Ricketts and Joseph Campbell up from Monterey to see and experience this place and this feeling, this stink and this nostalgia.

LEFT: Near Cedar Beach in Sitka, 1940. Nancy Ricketts's friend C. J. Mills lived in the house to the right while the family built the larger house on the left. RIGHT: Sitka's main dock, near modern-day Totem Square, 1940. *(Both photos taken with Nancy Ricketts's Brownie camera. Courtesy of Nancy Ricketts.)*

Of course, not all places are created equal. Sitka has almost a hundred inches of rain a year, about five times that of Monterey. So when it came time to go, Campbell and Ricketts went home. Calvin stayed. He became a noted conservationist and a leader in the community. Calvin stayed and ate the pickled herring and baked halibut.

Monterey of course had and still has the stink and the nostalgia, the smell of pizza and the rush of the tourist bus now, coffee bars and the lines for the aquarium. Still the great tidepool gathers water and the animals slither in and out as they did when Edward Weston, another influence on the *Grampus* crew, roamed the beaches with his large format camera, making his iconic black and white images. The eucalyptus trees drip with fragrant morning mist as they did when the experimental composer, John Cage lived with Jack Calvin's sister-in-law and re-imagined American music, and began his first thoughts leading him to incorporate Zen Buddhism into his philosophy of sound, nature and being.

In stories we all become larger than life. Even family anecdotes become untethered from history. So too places become sentimental in our retelling. But just as some places are natural settings for harbors or for battlefields, some places are natural settings for the gatherings of artists, storytellers and their characters: fishing ports almost always draw good food, young athletic people, and in such places there will always be parties, and there will be dancing and there will be good conversation and afterwards there will be sex. This is what brings the farmer in from the fields, and the fishermen home from sea. To observe the process of the earth is to understand sensuality and this sensuality is what brought the artists to Monterey and Sitka, as well as the salt spray and the nostalgia.

Literature of place is about love in the end. Steinbeck loved Pacific Grove and he loved Ed Ricketts—you can read that in every word he wrote about the man. Jack Calvin and Steinbeck didn't get along. There is no secret in this. Calvin was a doer and an adventurer. As a writer he was eclipsed by Steinbeck, but as an adventurer and conservationist he outstripped the Nobel Laureate. Jack loved the Alaskan wilderness, and in it he saw the world and the wildness that his old partner Ricketts saw just beyond the tidepool.

In Alaska Jack Calvin saw the world that was disappearing from Pacific Grove, and in Sitka, to this day, the poetry of the wild earth still lives on, its breath still blusters down the streets of salmonberry bushes, herring milt, and rusted iron roofing and yes, it sometimes stinks but never, ever has it lost its wild and sensual pull for those willing to be led.

Upstairs at Pacific Biological Laboratories. *(Photo by Ed Ricketts, Jr., circa 1947. Pat Hathaway Collection, CV#81-021-071.)*

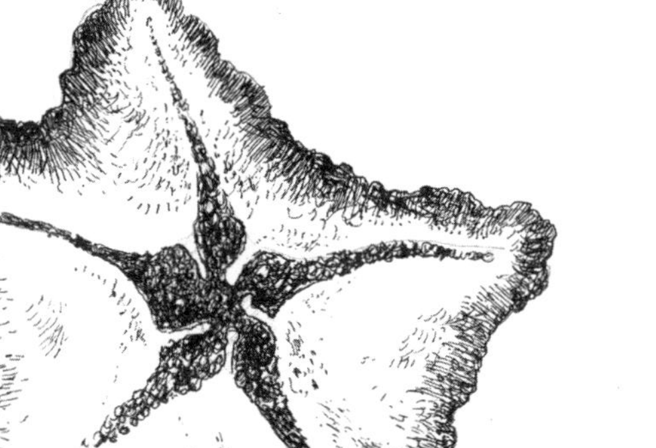

CONTRIBUTORS

Richard Astro is Provost Emeritus and Distinguished University Professor at Drexel University. Prior to coming to Drexel, he served as Provost at the University of Central Florida, as Dean of Arts and Sciences at Northeastern University, and as Professor and Chair of English at Oregon State University. He is the author of *John Steinbeck and Edward F. Ricketts: The Shaping of a Novelist* (1973), which was the first full-length book to deal with the Steinbeck-Ricketts connection. He has also written and/or edited books about Ernest Hemingway, Bernard Malamud, and the literature of New England, and numerous articles on American literature and higher education administration. He completed his doctorate in American literature at the University of Washington and then assumed his first teaching job at Oregon State where he met the infamous marine biologist, Joel Hedgpeth, who in turn introduced him to Ed Ricketts, Jr. The rest is history. As a respite from his university work, Richard served as a consultant on education programs for the Boston Red Sox, the Philadelphia Phillies, and most recently, for the New York Mets. He and his wife, Betty, have a daughter, Kelly; they split their time between Orlando and Vero Beach, Florida.

Donald Kohrs is Branch Library Specialist at the Harold A. Miller Library of Stanford University's Hopkins Marine Station in Pacific Grove, California. He lives in Santa Cruz and has degrees in biology and library science. His past research includes the history of Pacific Grove's Chautauqua program, the history of the Hopkins Seaside Laboratory, and the early years of Hopkins

Marine Station. His current project is uncovering the story behind the publication of *Between Pacific Tides*, the seminal work of marine biology by Edward F. Ricketts and Jack Calvin with illustrations by Ritchie Lovejoy.

David Lohse is a Research Specialist at the University of California Santa Cruz. He earned his PhD from the University of California Santa Barbara, studying mussel communities on rocky shores. He has taught marine ecology classes at several universities, and has been involved with PISCO (Partnership for Interdisciplinary Studies of Coastal Oceans) and MARINe (Multi-Agency Rocky Intertidal Network) since 1999. Through these programs, he has explored and documented patterns of stability and change in rocky shore communities ranging from Baja California, Mexico to Sitka, Alaska. In 2012 and 2018, he helped conduct biodiversity surveys in Sitka in areas visited by Ricketts and Calvin in 1932.

Melissa Miner is a Research Specialist at the University of California Santa Cruz, working remotely from Bellingham, Washington. She grew up in California, exploring the same tide pools where Ed Ricketts collected in Monterey Bay. She earned an MS from Moss Landing Marine Laboratories. She has worked on MARINe (Multi-Agency Rocky Intertidal Network)—a long-term shoreline monitoring project—for over twenty years and has helped to document changes that have occurred in rocky intertidal communities from Baja California, Mexico to Sitka, Alaska. In 2011, she helped establish long-term monitoring sites in Sitka in areas surveyed by Ricketts and Calvin in 1932, and was instrumental in the effort to compare their historical species records to current surveys.

Colleen Mondor is a writer and journalist who divides her time between Fairbanks, Alaska and the Pacific Northwest. She holds degrees in aviation, history, and northern studies and is the author of *The Map of My Dead Pilots: The Dangerous Game of Flying in Alaska,* among numerous other publications both online and in print.

John Pearse grew up in Tucson, Arizona, where he dreamed of becoming a desert biologist. However, after undergraduate work at the University of Chicago, he fled Chicago for California and marine biology, just like Ed Ricketts. In 1959, he arrived at Stanford University's Hopkins Marine Station in Pacific Grove, California, only a few blocks from Ed Ricketts's lab. And like Ricketts, he became enchanted with the rich life in the rocky intertidal and spent most of the rest of his life working along the shore of central California. In 1971 he joined the faculty of the University of California, Santa Cruz. There he taught courses in invertebrate zoology, intertidal biology, and kelp forest ecology, interweaving his teaching with his research in these same areas and inspiring countless undergraduate and graduate students. He and his wife, Vicki, teamed with her parents, Ralph and Mildred Buchsbaum, to revise their classic textbook *Animals Without Backbones* (1938), and also produced an expanded volume, *Living Invertebrates* (1987). Vicki's doctoral research was at Hopkins, and both she and John enjoyed decades of personal connections with

Hopkins Marine Station and with Hand, Abbott, and Davis. Their son, Devon Pearse, is an evolutionary geneticist working at the Southwest Fisheries Science Center in Santa Cruz. Because he lives just across Monterey Bay, the three were able to join forces on the Snad Rock study reported here in the "Legacy of a Naturalist" essay. After retiring in 1994, John launched LiMPETS (Long-term Monitoring Program and Experiential Training for Students), pioneering citizen science in which high-school students monitor the abundance of key intertidal organisms along the California coast. Now living and working in Pacific Grove next to Hopkins Marine Station, he continues to thrive in Ricketts's stomping grounds.

Peter Raimondi is a professor and chair in the Department of Ecology and Evolutionary Biology at the University of California Santa Cruz. He grew up in Tucson, Arizona, where a visit to the nearest tidepool required crossing the border into Mexico, and heading to the Gulf of California (Sea of Cortez), one of Ricketts's favorite locations to collect and explore. Originally trained as a student of philosophy at Northern Arizona University, Peter made the switch to science and received his PhD from the University of California Santa Barbara, studying barnacles in his familiar Gulf of California habitat. He joined UC Santa Cruz in 1995. During his tenure at UC Santa Cruz, he has engaged countless students of marine science through field courses both local and abroad. He has been a leader of the MARINe (Multi-Agency Rocky Intertidal Network) and PISCO (Partnership for Interdisciplinary Studies of Coastal Oceans) projects since their inception, and was part of the effort in 2011 to establish monitoring sites in Sitka in areas originally surveyed by Ricketts and Calvin in 1932.

Nancy Jane Ricketts is the eldest daughter of Edward F. Ricketts and Nan Makar Ricketts. Born in 1924, she lived in Pacific Grove with her parents until about 1932, and then moved to Carmel after her father's trip to Sitka, Alaska. She has lived in the Pacific Northwest, Alaska, and Baltimore, Maryland, working as an archivist, librarian, musician, writer, and a collector of her father's papers. She has lived in Sitka, Alaska for the last forty-six years.

Katharine A. Rodger spent over a decade researching the life and work of Ed Ricketts. She has edited two collections of his writings—*Renaissance Man of Cannery Row: The Life and Letters of Edward F. Ricketts* and *Breaking Through: Essays, Journals, and Travelogues of Edward F. Ricketts*. Katie has traveled around the country to speak about Ricketts, but she most enjoys visiting Sitka and Monterey. She currently teaches science and environmental writing in the University Writing Program at the University of California Davis.

Carolyn Servid is a writer and book designer who recently moved to Colorado's Western Slope after living in Sitka, Alaska for thirty-seven years. During that time, she co-founded and directed The Island Institute, a nonprofit organization whose programs focused on the nexus of story, place, and community. Her work with the Island Institute earned her the Alaska Governor's Humanities Award in 2001 and an honorary Doctorate of Humane Letters from the University of Alaska Southeast in 2008. Her books include a memoir, *Of Landscape and Longing: Finding a Home at the Water's Edge,* and three anthologies. Her essays have also appeared in various collections and literary journals. Among her book design credits are *Frank Mechau: Artist of Colorado* by Cile Bach, and reprints of *Square-Rigged* and *Fisherman 28,* two early novels by Jack Calvin.

Janice M. Straley holds an MS in Biological Oceanography and is Professor of Marine Biology at the University of Alaska Southeast Sitka Campus. For the past four decades she has conducted research on large whales, with a focus on whale interactions with commercial fishing gear. She has received numerous honors, including a Faculty Excellence in Research award in 2018, a University of Alaska Board of Regents Meritorious Service Award in 2013, and the Ocean Leadership Award for Excellence in Marine Science from the Alaska SeaLife Center in 2012. She resides along the wave swept coast at her home in Sitka, Alaska with her husband, John.

John Straley is the award-winning author of ten crime novels, including *The Woman Who Married A Bear* and *What is Time to a Pig?* Winner of the Shamus and the Spotted Owl awards for his detective fiction, John's novels are published all over the world. His nonfiction work, *Animal Nature,* is an entertaining biography of a local Sitka veterinarian. He has published five books of poetry: *The Rising and The Rain* and four seasonal collections of haiku, *Summer, Fall, Winter,* and *Spring.* The University of Alaska Fairbanks awarded him an honorary Doctorate in 2008, and he was appointed the Alaska Writer Laureate in 2006. He lives in a bright green house by Old Sitka Rocks with his wife Jan, a noted marine biologist and their wild and crazy dog, Dot.

WORKS CITED

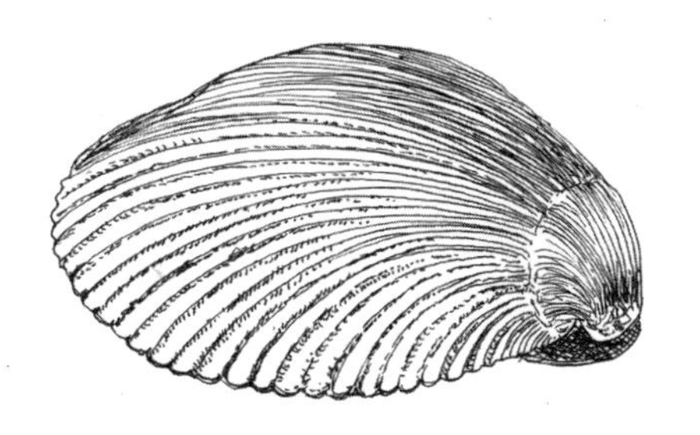

Ricketts, E. F. 2006. *Breaking Through: Essays, Journals and Travelogues of Edward F. Ricketts.* Katharine A. Rodger, ed. Oakland: University of California Press.

— 1925. "Foreword to the 1925 Biological Laboratories Catalog."

Ricketts, E. F. and Jack Calvin. 1948. *Between Pacific Tides: An Account of the Habits and Habitats of Some Five Hundred of the Common, Conspicuous Seashore Invertebrates of the Pacific Coast between Sitka, Alaska and Northern Mexico.* Rev. ed. Redwood City: Stanford University Press.

Hedgpeth, Joel. 1978. *The Outer Shores*. Eureka: Mad River Press.

Steinbeck, John. 1995. "About Ed Ricketts." *The Log from the Sea of Cortez*. New York: Penguin.

The *Grampus*

Bruckner, D.J.R. 1983. "Joseph Campbell: 70 Years of Making Connections." *New York Times,* 18 December, 1983: NYTimes.com.

Calvin, Jack. 1933. "Nakwasina Goes North: A Man, a Woman, and a Pup Cruise From Tacoma to Juneau in a 17-Foot Canoe." *The National Geographic Magazine,* July, 1933: 1-42.

Larsen, Stephen and Larsen, Robin. 2002. *Joseph Campbell: A Fire in the Mind*. Rochester: Inner Traditions.

Ricketts, E. F. 2020. "Notes and Observations, Mostly Ecological, Resulting from Northern Pacific Collecting Trips Chiefly in Southeastern Alaska, with Special Reference to Wave Shock as a Factor in Littoral Ecology." *Ed Ricketts: From Cannery Row to Sitka, Alaska*. Janice M. Straley, ed. Rev. ed. Sitka: Old Sitka Rocks Press.

Ricketts, E. F., Jack Calvin, Joel W. Hedgpeth, revised by David W. Phillips. 1985. *Between Pacific Tides*. Fifth ed. Redwood City: Stanford University Press.

Tamm, Eric Enno. 2004. *Beyond the Outer Shores*. New York: Four Walls Eight Windows.

RICKETTS, CALVIN, AND *BETWEEN PACIFIC TIDES*

Allee, W. C. 1923. "Studies in marine ecology, I and II," *Biology Bulletin*, 44: 157-253.

Campbell, Joseph and Moyers, Bill D. 1988. *The Power of Myth*. New York: Doubleday.

Ricketts, E. F. 2020. "Notes and Observations, Mostly Ecological, Resulting from Northern Pacific Collecting Trips Chiefly in Southeastern Alaska, with Special Reference to Wave Shock as a Factor in Littoral Ecology." *Ed Ricketts: From Cannery Row to Sitka, Alaska*. Janice M. Straley, ed. Rev. ed. Sitka: Old Sitka Rocks Press.

Ricketts, E. F. 1932. Letter of correspondence to Torsten Gislén, October 21, 1932. Torsten Gislén archive, Lund University Library. Sweden. (Keith Benson located and acknowledged the existence of Gislén's letters.)

Ricketts, E. F. 2002. *Renaissance Man of Cannery Row: The Life and Letters of Edward F. Ricketts*. Katharine A Rodger, ed. Tuscaloosa: University of Alabama Press.

Steinbeck, John, 1941. *Sea of Cortez: A Leisurely Journey of Travel and Research*. New York: Viking.

— *Log from the Sea of Cortez*. 1951. New York: Viking.

Beck MW, Keck KL Jr., Able KW, Childers DL, Eggleston DB, Gillanders BM, Halpern B, Hays CG, Hoshino K, Minello KJ, Orth RJ, Sheridan PF, Weinstein MP. 2001. "The identification, conservation, and management of estuarine and marine nurseries for fish and invertebrates." *BioScience* 51: 633-641.

Bell EC, Gosline JM. 1997. "Strategies for life in flow: tenacity, morphometry, and probability of dislodgement of two *Mytilus* species." *Marine Ecology Progress Series* 159: 197-208.

Blanchette CA. 1996. "Seasonal patterns of disturbance influence recruitment of the sea palm, *Postelsia palmaeformis*." *Journal of Experimental Marine Biology and Ecology* 197: 1-14.

— 1997. "Size and survival of intertidal plants in response to wave action: a case study with *Fucus gardneri*." *Ecology* 78: 1563–1578.

Blanchette CA, Helmuth B, Gaines SD. 2007. "Spatial patterns of growth in the mussel, *Mytilus californianus*, across a major oceanographic and biogeographic boundary at Point Conception, California, USA." *Journal of Experimental Marine Biology and Ecology* 340 (2): 126-148.

Boulding EG, Holst M, Pilon V. 1999. "Changes in selection on gastropod shell size and thickness with wave exposure on Northeastern Pacific shores." *Journal of Experimental Marine Biology and Ecology* 232: 217-239.

Burrows MT, Harvey R, Robb L. 2008. "Wave exposure indices from digital coastlines and the prediction of rocky shore community structure." *Marine Ecology Progress Series* 353: 1-12.

Bustamante RH, Branch GM. 1996. "Large scale patterns and trophic structure of southern African rocky shores: The roles of geographic variation and wave exposure." *Journal of Biogeography* 23 (3): 339-351.

D'Antonio, CM. 1986. "Role of sand in the domination of hard substrata by the intertidal alga *Rhodomela larix*." *Marine Ecology Progress Series* 27: 263-275.

Dayton PK. 1971. "Competition, disturbance, and community organization: the provision and subsequent utilization of space in a rocky intertidal community." *Ecological Monographs* 41: 351-389.

— 1973. "Dispersion, dispersal, and persistence of the annual intertidal alga, *Postelsia palmaeformis Ruprecht*." *Ecology* 54: 433-438.

— 1975. "Experimental evaluation of ecological dominance in a rocky intertidal algal community." *Ecological Monographs* 45: 137-159.

Denny MW. 1987. "Lift as a mechanism of patch initiation in mussel beds." *Journal of Experimental Marine Biology and Ecology* 113: 231-245.

— 1995. "Predicting physical disturbance—mechanistic approaches to the study of survivorship on wave-swept shores." *Ecological Monographs* 65 (4): 371-418.

Denny MW, Helmuth B, Leonard GH, Harley CDG, Hunt LJH, Nelson EK. 2004. "Quantifying scale in ecology: lessons from a wave-swept shore." *Ecological Monographs* 74: 513-532.

Fisher WK. 1930. "Asteroidea of the North Pacific and Adjacent Waters. Part 3. Forcipulata (Concluded)" *United States National Museum, Bulletin 76*: Washington, D.C.

Flattely FW, Walton L. 1922. *The Biology of the Seashore*. New York: Macmillan Co.

Foster MS, Schiel DR. 1985. "The ecology of giant kelp forests in California: a community profile." *US Fish and Wildlife Biological Report* 85 (7.2).

Gaylord B, Blanchette CA, Denny MW. 1994. "Mechanical consequences of size in wave swept algae." *Ecological Monographs* 64: 287-313.

Graham MH, Halpern BS, Carr MH. 2008. "Diversity and dynamics of California subtidal kelp forests: disentangling trophic interactions from habitat associations." *Food Webs and the Dynamics of Marine Reefs*, T McClanahan and GM Branch, eds. Oxford: Oxford University Press.

Guichard F, Halpin PM, Allison GW, Lubchenco J, Menge BA. 2003. "Mussel Disturbance Dynamics: Signatures of Oceanographic Forcing from Local Interactions." *American Naturalist* 161: 889-904.

Hagerman G, Bedard R. 2004. "Offshore wave power in the US: Environmental issues." *Electric Research Power Institute Report*: E2I Global EPRI-007-US.

Hayne KJR, Palmer RA. 2013. "Intertidal sea stars *(Pisaster ochraceus)* alter body shape in response to wave action." *The Journal of Experimental Biology* 216: 1717-1725.

Johnson ME, Snook HJ. 1927. *Seashore Animals of the Pacific Coast*. New York: Macmillan Co.

Largier JL, Behrens MD, Robart M. 2008. "The Potential Impact of WEC Development on Nearshore and Shoreline Environments through a Reduction in Nearshore Wave Energy." In *Developing Wave Energy In Coastal California: Potential Socio-Economic and Environmental Effects* (pp. 57-81). California Energy Commission, PIER Energy-Related Environmental Research Program & California Ocean Protection Council CEC-500-2008-083.

Leigh EG, Paine RT, Quinn JF, Suchanek TH. 1987. "Wave energy and intertidal productivity." *Proceedings of the National Academy of Science* 84: 1314-1318.

Lindegarth M, Gamfeldt L. 2005. "Comparing categorical and continuous ecological analyses: effects of 'wave exposure' on rocky shores." *Ecology* 86: 1346-1357.

Littler MM, Martz DR, Littler DS. 1983. "Effects of recurrent sand deposition on rocky intertidal organisms: importance of substrate heterogeneity in a fluctuating environment." *Marine Ecology Progress Series* 11: 129-139.

Lohse DP, Gaddam RN, Raimondi PT. 2008. "Predicted effects of Wave Energy Conversion on communities in the nearshore environment." In *Developing Wave Energy In Coastal California: Potential Socio-Economic and Environmental Effects* (pp. 83-109). California Energy Commission, PIER Energy-Related Environmental Research Program & California Ocean Protection Council CEC-500-2008-083.

McGuinness KA. 1987. "Disturbance and organisms on boulders I. Patterns in the environment and the community." *Oecologia* 71: 409-419.

— 1987. "Disturbance and organisms on boulders II. Causes of patterns in diversity and abundance." *Oecologia* 71: 420-430.

Menge BA. 1976. "Organization of the New England rocky intertidal community: role of predation, competition and environmental heterogeneity." *Ecological Monographs* 46: 355-393.

— 1978. "Predation intensity in a rocky intertidal community." *Oecologia* 34: 1-16.

Menge BA, Allison GW, Blanchette CA, Farrell TM, Olson AM, Turner TA, van Tamelen P. 2005. "Stasis or kinesis? Hidden dynamics of a rocky intertidal macrophyte mosaic revealed by a spatially explicit approach." *Journal of Experimental Marine Biology and Ecology* 314: 3-39.

Menge BA, Berlow EL, Blanchette CA, Navarrete SA, Yamada SB. 1994. "The keystone species concept: variation in interaction strength in a rocky intertidal habitat." *Ecological Monographs* 64: 249-286.

Menge BA, Farrell TM, Olson AM, van Tamelen P, Turner T. 1993. "Algal recruitment and the maintenance of a plant mosaic in the low intertidal region on the Oregon coast." *Journal of Experimental Marine Biology and Ecology* 170: 91-116.

Nielsen KJ. 2001. "Bottom-up and top-down forces in tide pools: test of a food chain model in an intertidal community." *Ecological Monographs* 71: 187-217.

— 2003. "Nutrient loading and consumer: agents of change in open-coast macrophyte assemblages." *Proceedings of the National Academy of Sciences* 100: 7660-7665.

Nielsen KJ, Blanchette CA, Menge BA, Lubchenco JL. 2006. "Physiological snapshots reflect ecological performance of the sea palm, *Postelsia palmaeformis* (Phaeophyceae) across intertidal elevation and exposure gradients." *Journal of Phycology* 42: 548-559.

Paine RT. 1966. "Food web complexity and species diversity." *American Naturalist* 100: 65-75.

— 1974. "Intertidal community structure: experimental studies on the relationship between a dominant competitor and its principal predator." *Oecologia* 15: 93-120.

— 1979. "Disaster, catastrophe, and local persistence of the sea palm *Postelsia palmaeformis.*" *Science* 205: 685-687.

— 1988. "Habitat suitability and local population persistence of the sea palm *Postelsia palmaeformis*." *Ecology* 69: 1787-1794.

Paine RT, Levin SA. 1981. "Intertidal landscapes: disturbance and the dynamics of pattern." *Ecological Monographs* 51: 145–178.

Reed DC, Rassweiler A, Carr MH, Cavanaugh KC, Malone DP, Siegel DA. 2011. "Wave disturbance overwhelms top-down and bottom-up control of primary production in California kelp forests." *Ecology* 92: 2108-2116.

Ricketts EF, Calvin J. 1939. *Between Pacific Tides*. Stanford: Stanford University Press.

Robles C, Desharnais R. 2002. "History and current development of a paradigm of predation in rocky intertidal communities." *Ecology* 83: 1521-1536.

Robles C, Garza C, Desharnais RA, Donahue MJ. 2010. "Landscape patterns in boundary intensity: a case study of mussel beds." *Landscape Ecology* 25: 745-759.

Rosenthal RJ, Clarke WD, Dayton PK. 1974. "Ecology and natural history of a stand of giant kelp, *Macrocystis pyrifera*, off Del Mar, California." *United States National Marine Fisheries Service Fishery Bulletin* 72: 670-684.

Schiel DR, Foster MS. 2015. *The Biology and Ecology of Giant Kelp Forests*. Oakland: University of California Press.

Schoch GC, Dethier M. 1996. "Scaling up: the statistical linkage between organismal abundance and geomorphology on rocky intertidal shorelines." *Journal of Experimental Marine Biology and Ecology* 201-37-72.

Sebens KP. 2002. "Energetic constraints, size gradients, and size limits in benthic marine invertebrates." *Integrative and Comparative Biology* 42: 853-861.

Shanks AL, Wright WG. 1986. "Adding teeth to wave action: the destructive effects of wave-borne rocks on intertidal organisms." *Oecologia* 69: 420-428.

Sousa WP. 1979. "Disturbance in marine intertidal boulder fields: the non-equilibrium maintenance of species diversity." *Ecology* 60: 1225-1239.

Taylor PR, Littler MM. 1982. "The roles of compensatory mortality, physical disturbance, and substrate retention in the development and organization of a sand-influenced, rocky-intertidal community." *Ecology* 63: 135-146.

Verrill AE, Smith SI. 1874. "Invertebrate animals of Vineyard Sound and adjacent waters with an account of the physical features of the region." *Report of Professor S.F. Baird, Commissoner of Fish and Fisheries, on the condition of the sea-fisheries of the south coast of New England in 1871 and 1872.* Washington DC: Government Printing Office.

Zabin CJ, Danner EM, Baumgartner EP, Spafford D, Miller KA, Pearse JS. 2012. "A comparison of intertidal species richness and composition between central California and Oahu." *Marine Ecology* 2012: 1-16.

SUGGESTED READINGS

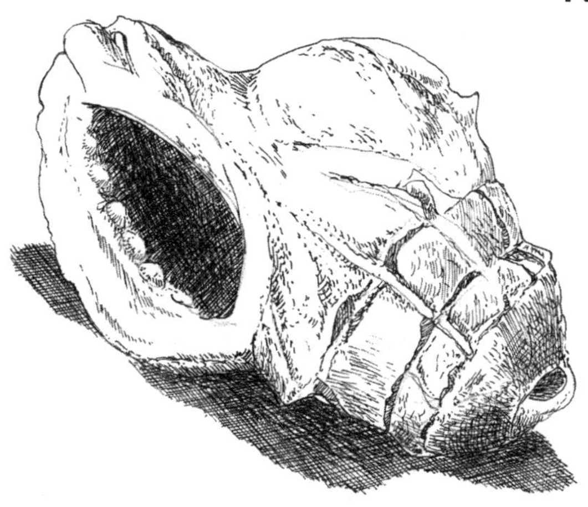

Astro, Richard. 1976. *Edward F. Ricketts. (Western Writers Series No. 21)* Boise: Boise State University.

Baldwin, Aaron, and Paul Norwood. 2015. *Common Sea Life of Southeastern Alaska: A Field Guide.* 15 Aug., 2015. Web: https://www.naturebob.com. Select LINKS AND RESOURCES, scroll to HABITATS, listing #7: *Intertidal.*

Buchsbaum, Ralph, Mildred Buchsbaum, John Pearse, and Vicki Pearse. 1987. *Animals Without Backbones: An Introduction to the Invertebrates.* Chicago: University of Chicago Press.

Calvin, Jack. 1933. "Nakwasina Goes North: A Man, a Woman and a Pup Cruise From Tacoma to Juneau in a 17-Foot Canoe." *The National Geographic Magazine,* July 1933: 1-42.

Campbell, Joseph with Bill Moyers. 1988. *The Power of Myth.* New York: Doubleday.

Hedgpeth Joel. 1978. *The Outer Shores Part 1.* Eureka: Mad River Press, Inc.

— 1978. *The Outer Shores Part 2.* Eureka: Mad River Press, Inc.

Kozloff, Eugene. 1993. *Seashore Life of the Northern Pacific Coast. An Illustrated Guide to Northern California, Oregon, Washington, and British Columbia.* Seattle: University of Washington Press.

Lamb, Andy, and Benard Hanby. 2005. *Marine Life of the Pacific Northwest: A Photographic Encyclopedia of Invertebrates, Seaweeds and Selected Fishes.* Pender Harbour: Harbour Publishing.

Lannoo, Michael. 2010. *Leopold's Shack and Ricketts's Lab: The Emergence of Environmentalism.* Oakland: University of California Press.

Larsen, Stephen and Robin Larsen. 2002. *Joseph Campbell: A Fire in the Mind.* Rochester: Inner Traditions.

Levings, Colin D. 2020. "Edward Flanders Ricketts and the marine ecology of the inner coast habitats of British Columbia, Canada." *Archives of Natural History,* 47.1 (2020): 115-123. Edinburgh: Edinburgh University Press.

Lindeberg, Mandy, and Sandra Lindstrom. 2010. *Field Guide to Seaweeds of Alaska.* Fairbanks: Alaska Sea Grant College Program, University of Alaska Fairbanks.

Miner, Melissa, Pete Raimondi, Lisa Busch, Janice Straley, Taylor White, Rani Gaddam, Marnie Chapman, and Victoria O'Connell. 2013. *Between Pacific Tides: Revisiting Historical Surveys of Sitka through Ricketts, Calvin and Ahlgren.* Anchorage: North Pacific Research Board Project 1115 Final Report. Aug 2013. Web: http://projects.nprb.org. Select CORE PROGRAM, then PROJECT SEARCH AND DATABASE. Enter **1115** in the Search window below.

Moyers, Bill. 1988. *Joseph Campbell and the Power of Myth.* PBS TV Mini Series.

Pearse, Vicki, John Pearse, Mildred Buchsbaum, and Ralph Buchsbaum. 1987. *Living Invertebrates.* Pacific Grove: The Boxwood Press.

Ricketts, Edward, and Jack Calvin. 1948. *Between Pacific Tides: An Account of the Habits and Habitats of Some Five Hundred of the Common, Conspicuous Seashore Invertebrates of the Pacific Coast between Sitka, Alaska and Northern Mexico,* Rev. ed. Redwood City: Stanford University Press.

Ricketts, Edward, Jack Calvin and Joel Hedgpeth, revised by David Phillips. 1985. *Between Pacific Tides,* Fifth ed. Redwood City: Stanford University Press.

Ricketts, Edward. 2002. *Renaissance Man of Cannery Row: The Life and Letters of Edward F. Ricketts,* edited by Katharine A. Rodger. Tuscaloosa: University of Alabama Press.

— 2006. *Breaking Through: Essays, Journals, and Travelogues of Edward F. Ricketts,* edited by Katharine A. Rodger. Oakland: University of California Press.

Steinbeck, John, and Edward Ricketts. 1941. *The Sea of Cortez*. New York City: Viking Press.

Steinbeck, John. 2002. *Cannery Row* (Steinbeck Centennial Edition). New York City: Penguin Books.

Straley, John. 1995. *The Curious Eat Themselves*. New York City: Soho Crime.
— 2014. *The Big Both Ways*. New York City: Soho Crime.

— 2008. *The Rising and the Rain*. Fairbanks: University of Alaska Press.

Tamm, Eric. 2004. *Beyond the Outer Shores.* New York City: Four Walls Eight Windows.

Websites

Intertidal Long-Term Monitoring: http://www.pacificrockyintertidal.org

More on Ed Ricketts: https://seaside.stanford.edu/ricketts

Encounters North Radio Podcast:
http://www.encountersnorth.org/
https://www.encountersnorth.org/listen-index

North Pacific Research Board:
http://www.nprb.org/news/detail/between-pacific-tides-project-on-the-radio

SPECIES LIST

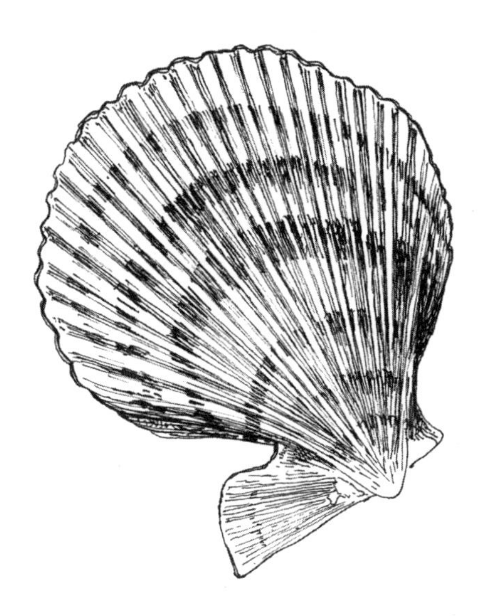

Comparison of Sitka species described in 1932 survey by Ricketts and Calvin (R/C) to re-surveys done in 2012 and 2018 by UC Santa Cruz (UCSC) biodiversity group (Miner et al. 2013). Marnie Chapman, Univeristy of Alaska Southeast Professor of Biology, compiled the species list from the 1932 "Wave Shock Essay." John and Vicki Pearse assisted with veryifying species listed for the comparison.

Table 1. Comparison of Sitka species described in 1932 survey by Ricketts and Calvin (R/C) to re-surveys done in 2012 and 2018 by UC Santa Cruz (UCSC) biodiversity group.

Phylum	R/C Name (1932)	Current Name (if different)	Habitat where R/C found	R/C Site(s)	UCSC 2012	UCSC 2018**	Included in Comparison?
Porifera							
	Reneira cineria	Haliclona sp.	rocky reef	KIO, KII	KIO, KII	CB, SR	Yes
Cnidaria							
	Abietinaria		rocky reef	KIO	KIO, KII	KII	Yes
	Bunodactis "artemisia form"	Anthopleura artemisia	coarse sand	?	KII, CB, SR	KII, CB, SR	Yes
	Bunodactis "elegantissima form" *	Anthopleura elegantissima	rocky reef	KII			Yes
	Bunodactis "xanthogrammica form"	Anthopleura xanthogrammica	rocky reef	KIO	KIO, KII, CB	KII, CB	Yes
	Charisea? *	Charisea saxicola	mud flat/rock mix, rocky reef	JB, KIO, KII			Yes
	Eudendrium *		rocky reef	KIO			Yes
	Gersemia rubiformis		rocky reef	KIO	KII		Yes
	Gonionemus vertens		eel grass beds	?			No
	Gonothyraea		floats	?			No
	Metridium	Metridium farcimen	piles and floats	?	KIO, KII, SR	KII, CB	Yes
	Obelia longissima		eel grass beds	?			No
	Polyorchis	Polyorchis penicillatus	eel grass beds	CB			No
	Stylantheca porphyra	Stylantheca spp	rocky reef	KIO	KIO		Yes
	Urticina		rocky reef	CB	KIO	CB	Yes

CB=Cape Baranof/Pirate Cove
JB=Jametown Bay (not sampled by UCSC)
KII=Kayak Islands Inner Coast
KIO=Kayak Islands Outer Coast (not sampled 2018)
SR=Sage Rock (not sampled by R/C),
?=Site Unspecified

Phylum	R/C Name (1932)	Current Name (if different)	Habitat where R/C found	R/C Site(s)	UCSC 2012	UCSC 2018**	Included in Comparison?
Platyhelminthes							
	giant flatworms			?			No
Nemertea							
	nemerteans		mud flat	JB			No
Annelida							
	Arenicola		mud flat	JB			No
	Glycera		mud flat/rock mix	JB			No
Echiura							
	Echiurus pallassii	Echiurus echiurus	mud flat	JB			No
Sipuncula							
	Physcosoma agassizii	Phascolosoma agassizii	mud flat/rock mix	CB			No
Mollusca							
	Acmaea	Acmaea mitra?	rocky reef	KII	KII, KIO, CB	KII, CB	Yes
	Calliostoma costatum	Calliostoma ligatum	rocky reef/kelp bed	KIO	KIO, CB	KII, CB	Yes
	Cardium	Clinocardium nuttallii	coarse sand	?			No
	Cryptochiton	Cryptochiton stelleri	rocky reef	CB	KII, CB	CB	Yes
	Dialula	Diaulula sandiegensis	rocky reef/kelp bed	KIO	KII, KIO		Yes
	Haliotis kamtschatkana *		rocky reef	CB			Yes
	Katherina	Katharina tunicata	rocky reef	KIO	KII, KIO, CB, SR	KII, CB, SR	Yes
	Mytilus californicanus	Mytilus californianus	rocky reef	KIO	KIO, CB	KII, CB	Yes
	Polinices	Neverita lewisii	mud flat	JB			No

Phylum	R/C Name (1932)	Current Name (if different)	Habitat where R/C found	R/C Site(s)	UCSC 2012	UCSC 2018**	Included in Comparison?
	Saxidomus	Saxidomus gigantea	mud flat	JB			No
	Tagula?	Tegula pulligo?	rocky reef	KIO		KII, CB	No
	Thais emarginata	Nucella ostrina	rocky reef	KIO	KIO, CB, SR	KII, CB, SR	Yes
	Thais lamellosa *	Nucella lamellosa	rocky reef	CB			Yes
Arthropoda							
	Cryptolithodes		rocky reef	CB	CB		Yes
	giant Balanus	Balanus nubilus?	rocky reef	KIO	KIO		Yes
	Haplogaster mertensii *	Hapalogaster mertensii	rocky reef	CB			Yes
	Hippolyte californiensis		eel grass beds	?			No
	Lophopanopeus bellus *		rocky reef	CB			Yes
	Mitella	Pollicipes polymerus	rocky reef	KIO	KIO	KII	Yes
	Peltogaster		parasite on crabs	KIO			No
	Pentidotea	Pentidotea resecata?	eel grass beds				No
	Pinnixa schmitti	Scleroplax schmitti	mud flat, commensal with terebellids	JB, CB			No
	Sacculina		parasite on crabs	KIO			No
	Spirontocaris camtschaticos		eel grass beds	JB			No
	Spirontocaris cristala		eel grass beds	JB			No
	Spirontocaris paludicala		eel grass beds	JB			No
Bryozoa							
	Flustra	Dendrobeania lichenoides?	rocky reef	KIO	KII, KIO, CB		Yes

Phylum	R/C Name (1932)	Current Name (if different)	Habitat where R/C found	R/C Site(s)	UCSC 2012	UCSC 2018**	Included in Comparison?
Echinodermata							
	Cucumaria miniata *		rocky reef	?			Yes
	Evasterias troschelii		mud flat/rock mix, rocky reef	JB,CB	KII, KIO, CB, SR	KII, CB, SR	Yes
	Mediaster aequalis		subtidal (from dredge)				No
	Ophiopholis		under boulders	?			No
	Pisaster ochraceus		rocky reef	KIO, CB	KII, KIO, CB, SR	KII, CB, SR	Yes
	Pycnopodia helianthoides		rocky reef	CB	KII, CB		Yes
	sand dollars	Dendraster excentricus	coarse sand	?			No
	Stichopus	Parastichopus californicus?	subtidal (from dredge)	?			No
	Strongylocentrotus drobachiensis		mud flat/rock mix, rocky reef	JB, KIO	KII, KIO, SR	KII, SR	Yes
	Strongylocentrotus franciscanus	Mesocentrotus franciscanus	rocky reef	KIO, CB	KIO	KII	Yes

	R/C	UCSC 2012	UCSC 2018
Total # species/species groups recorded:	**59**	**24**	**18**
Total # species/species groups appropriate for comparison:	**33**	**33**	**30**
% of R/C species/species groups found in UCSC surveys		**73%**	**60%**

*Not found in resurveys of ROCKY intertidal Sitka sites done by UCSC in 2012 (additional species were not found, but occur in habitat not surveyed)
**note that KIO, where many "exposed" species occur, was not surveyed in 2018

ACKNOWLEDGMENTS

This book could not have been possible without the support of Nancy Ricketts and Ed Ricketts Jr., who both saw the value in making the "Wave Shock Essay" available to a wider audience. Ed Jr. generously provided insights about the connection his father had with Sitka and the ideas he explored within the "Wave Shock Essay." Nancy Ricketts's patience and understanding in allowing me into her home to scan her photographs was greatly appreciated. Her insightful observations comparing Sitka today and Pacific Grove of the 1920s and 1930s were invaluable. I cherish the conversations I have had with her over the past five years and will forever value her friendship. I also thank Larry Frisch for suggesting that the "Wave Shock Essay" be published. I am very grateful to Pat Hathaway from Pacific Grove, CA who graciously provided high resolution images of some of the photographs presented within the essays.

Each of the six essayists (and co-authors) in this book readily accepted my challenge—to write an essay to provide background and context to Ed Ricketts's "Wave Shock Essay." The lead authors of the original essays met in Sitka as a group, some traveling long distances, to read their own essay aloud, listen to the others and make necessary changes to align content if needed. The sixth and newest essay gives a broader understanding of the influence of the *Grampus* voyage on the writings of Ed Ricketts and Joseph Campbell. I am indebted to these authors for their enthusiasm and support of my ideas and for creating a book that has far surpassed my expectations.

Marnie Chapman is passionate about having young students discover life inside a tidepool and under rocks on the beach. Marnie was responsible for the first comparison between the species listed in the "Wave Shock Essay" and the re-survey of the same locations eighty years later. Vicki and John Pearse reviewed and corrected the species list for taxonomic accuracy. They gave us permission to use the photograph taken by Vicki's father of Ed Ricketts holding the squid.

Liz McKenzie's cover-to-cover precise proofreading of this revised edition was invaluable.

Artist Norman Campbell has been a full partner in the creation of this book. From the cover to the many illustrations throughout the book, he captured the visual beauty, emotional impact, and scientific detail of this rugged coastline.

Finally, Carolyn Servid has created a visual masterpiece with every detail carefully and painstakingly designed for clarity and appearance. I thank her for those efforts.

Janice M. Straley
Old Sitka Rocks
Sitka, Alaska | May, 2020

ABOUT THE ARTIST

Oregon-born Norman Campbell moved to Southeast Alaska in 1982. Alaska has been the major source of inspiration for his work since his arrival. Its small intimate islands, its trees, and its spirit have been an abiding wellspring of ideas and images. Campbell works primarily with pen and ink on paper. He considers his work to be invented landscape and uses it to speak about his concerns, his vision, and his loves. He makes his home in Sitka, Alaska.

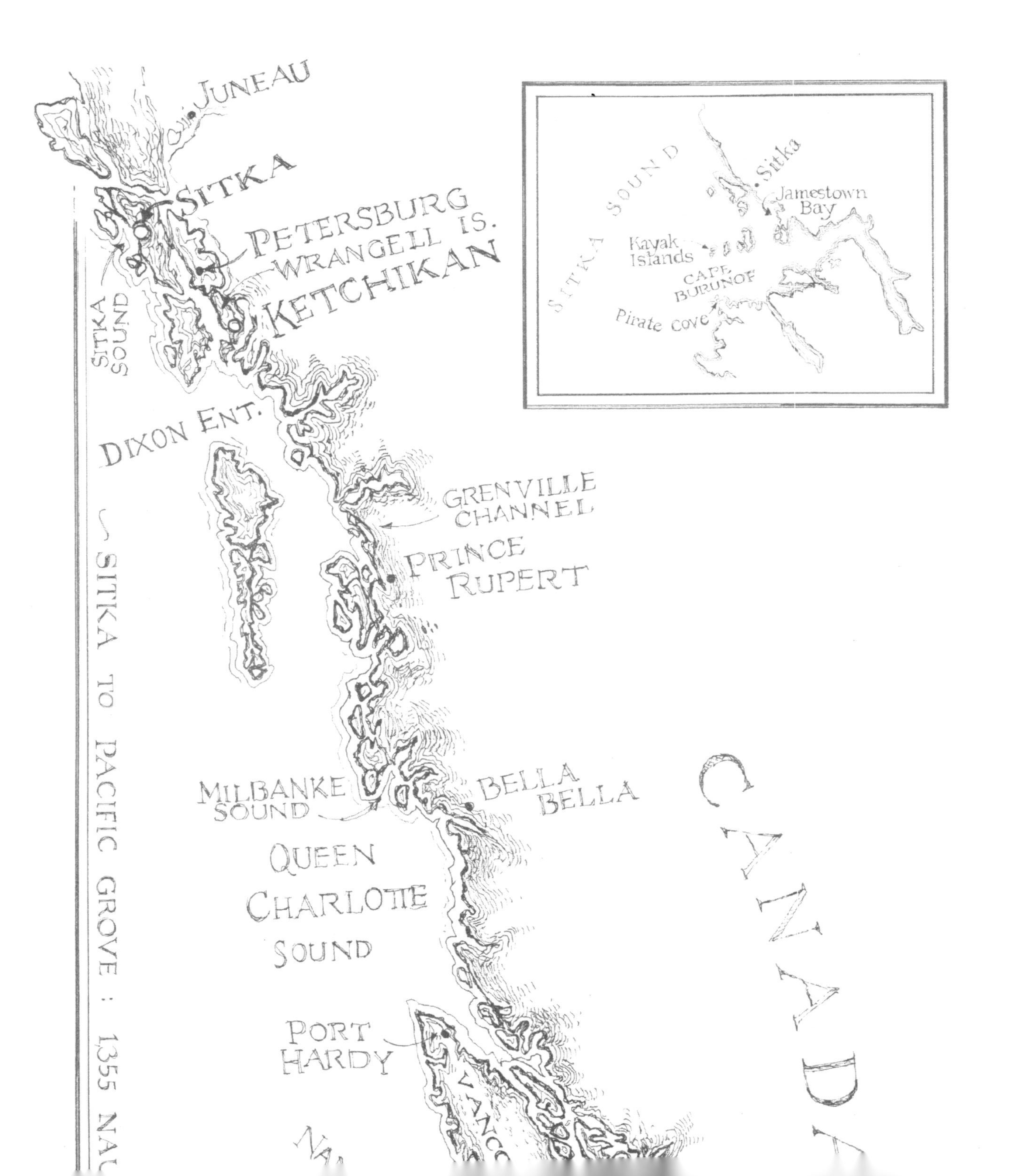
JUNEAU
SITKA
PETERSBURG
WRANGELL IS.
KETCHIKAN
SITKA SOUND
Sitka
Jamestown Bay
SITKA SOUND
Kayak Islands
CAPE BURUNOF
Pirate Cove
DIXON ENT.
GRENVILLE CHANNEL
PRINCE RUPERT
MILBANKE SOUND
BELLA BELLA
QUEEN CHARLOTTE SOUND
PORT HARDY
SITKA TO PACIFIC GROVE : 1,355